The Peopling of the Earth

The Peopling of the Earth

A commentary on
archaic records in
The Secret Doctrine

by

Geoffrey Barborka

THE THEOSOPHICAL PUBLISHING HOUSE
Wheaton, Ill., U.S.A.
Madras, India / London, England

Original edition 1975, published by The Theosophical Publishing House, Wheaton, Illinois, a department of The Theosophical Society in America

Library of Congress Catalog Card Number: 75-4243

ISBN: 0-8356-0221-4

Printed in the
United States of America

CONTENTS

DIAGRAMS

PREFACE

In 1969 the most spectacular event of the twentieth century was consummated: an event never before achieved during the known history of the world. It represented the culmination of years of scientific and technological endeavor —the sending of spacecraft outside the Earth's field of attraction. One spacecraft after another was launched; finally, on July 16, 1969, Apollo 11, bearing three American astronauts, was sent out into the lunar orbit. On July 20, these astronauts lowered a ladder onto the surface of the moon. The historic moment arrived on that date at 10:56 p.m. EDT at 23.49°E. and 0.67°N. As millions of eager viewers watched their television sets, Neil A. Armstrong descended the ladder and actually set foot on the moon. Later, his companion, Edwin E. Aldrin, also descended the ladder, and both astronauts proceeded to walk on the moon and to set up scientific equipment. Then they gathered lunar rocks and dust, raised an American flag, and took pictures. These features added to the thrill and excitement shared by people all over the world. After a stay of twenty-one and one-half hours, the two astronauts ascended the ladder and reentered their spacecraft for the return journey to the earth, thus successfully completing this historic venture.

In due time scientists proclaimed their findings, one of which was in full agreement with a statement made by those who had presented the teachings of the Ancient Wisdom to the Western world: there was no bacterial life on the moon. Consequently, it was declared to be a dead planet, a satellite. As phrased in *The Secret Doctrine*:

> . . . Planetary Chains have their "Days" and their "Nights"—i.e., periods of activity or life, and of inertia or death—and behave in heaven as men do on Earth: they generate their likes, get old, and become personally extinct, their spiritual principles only living in their progeny as a survival of themselves.[1]

> . . . the Law for the birth, growth, and decay of everything in the Kosmos, from the Sun to the glow-worm in the grass is One. It is an everlasting work of perfection with every new appearance, but the Substance-Matter and Forces are all one and the same. But this LAW acts on every planet through minor and varying laws.[2]

Why should an account of the peopling of the Earth be preceded by reference to a landing on the Moon? For the simple reason that the philosophy presented in *The Secret Doctrine* regards the Moon as the parent of the Earth and holds that, many, many millions of years ago, the present inhabitants of the Earth dwelt on the Moon; hence they are termed Lunar Monads. This statement is based on a passage in *The Secret Doctrine*:

> . . . who are the "Lunar Monads"? . . . they are the Monads who, having ended their life-cycle on the lunar chain, which is inferior to the terrestrial chain, have incarnated on this one.[3]

The present work not only bases its account on the volumes entitled *The Secret Doctrine,* but is written as a commentary on the Stanzas of Dzyan (Nos. I to IV) which were presented in the second volume of H. P. Blavatsky's great work and are regarded as the second series of the Stanzas—an archaic manuscript.

---

1. Blavatsky, H. P., *The Secret Doctrine,* I:154-5; I:209-10, 6-vol. ed.; I:178, 3rd ed. (Subsequent references to this work will be designated SD, except where *ibid.* is used to indicate the same reference source.)

2. Ibid., I:144-5; I:200, 6-vol. ed.; I:168, 3rd ed.

3. Ibid., I:179; I:231, 6-vol. ed.; I:202, 3rd ed.

In the seventh śloka of Stanza II, in this second series of the Stanzas of Dzyan, there is a very intriguing sentence. This is especially so in view of the fact that no commentary was supplied by H. P. Blavatsky to explain this particular sentence. Usually when a portion of a stanza or even a śloka (or verse) seems to be difficult of comprehension, a commentary has been added which offers enlightenment. Seeking for an elucidation of śloka 7, a systematic study of Stanzas I to IV was undertaken, resulting in the manuscript entitled *The Peopling of the Earth.*

For those who are undertaking a study of the Stanzas of Dzyan such as the one here presented, it should be made clear that the Esoteric Philosophy is based upon deduction, and this is the method adopted in this book. When presenting this philosophy to Mr. A. P. Sinnett in 1882, the Mahatma K. H. stated:

> . . . to show you how exact a science is occultism let me tell you that the means we avail ourselves of are all laid down for us in a code as old as humanity to the minutest detail, but everyone of us has to begin from the beginning, not from the end. Our laws are as immutable as those of Nature . . . we build our philosophy upon experiment and deduction.[4]

Mention should be made of the concept which is herein brought forward concerning systems of worlds, and the statement that the Earth system of worlds, termed the Earth-Chain, consists of seven globes. It is based upon this proposition:

> Everything in the metaphysical as in the physical Universe is septenary. Hence every sidereal body, every planet, whether visible or invisible, is credited with six companion globes. . . . These globes are formed by a process which the Occultists call the "rebirth of planetary chains."[5]

---

4. Barker, A. T., ed., *The Mahatma Letters to A. P. Sinnett,* p. 144, 2nd ed.; pp 140-1, 3rd ed. (Subsequent references to this work will be designated ML, except where *ibid.* is used to indicate the same reference source.)

5. SD I:158-9; I:213, 6-vol. ed.; I:182, 3rd ed.

Then, too, the concept that the Earth system of seven worlds is situated on a plane of activity superior to that in which the Moon system functioned, is thus stated:

> Every such chain of worlds is the progeny and creation of another, *lower* and *dead* chain—its reincarnation, so to say.[6]

Since this work is written as a commentary on the Stanzas of Dzyan published in *The Secret Doctrine,* it is understandable that the author is indebted to that work for all the material furnished herein. It is offered in the hope that it will be of assistance to readers in understanding the symbols and cryptic expressions recorded in those archaic scriptures.

It is fitting that an expression of appreciation should be rendered to Joy Mills and Virginia Hanson of the headquarters staff of The Theosophical Society in America, at Wheaton, Illinois, for their enthusiastic acceptance of the manuscript for publication by the Theosophical Publishing House, so that this work may appear during the centenary World Congress of The Theosophical Society in New York in 1875.

—GEOFFREY A. BARBORKA

Ojai, California

6. Ibid. I:152; I:207, 6-vol. ed; 1:176, 3rd ed.

1.

WHERE DO PEOPLE COME FROM?

HAVE you ever wondered: Where do all the people in the world come from? Omitting the obvious biological explanation, "From one's parents," the question provides an interesting topic for discussion. Thoughtful individuals are deeply concerned about the present increase in population; some speculate that there will be food shortages; others dwell upon catastrophic conditions which could result from overcrowding; still others are fearful about the pollution of the atmosphere and water, and are concerned about insecticide spraying which is used to insure good crops; while city officials are troubled by problems of waste disposal, and so forth. Musing on the question, George Macdonald's little gem came to mind:

> Where do you come from, baby dear?
> Out of the everywhere into the here.

There is a great deal of esoteric truth in these lines although, of course, they do not give a satisfactory answer to those who are worried about the increase in the world's population. One could list various theories and opinions to explain this increase, but it would be to no real purpose. Instead of speculating on the subject, let us offer a passage from *The Secret Doctrine* which, if accepted and thoroughly understood, would provide a definite answer to the question of man's origin. But before quoting the passage, the meaning of a word which helps to make the answer explicit should be understood. The term is "monads." It is clearly defined in this quotation:

> "Pilgrim" is the appellation given to our *Monad* (the two in one) during its cycle of incarnations. It is the only immortal and eternal principle in us, being an indivisible part of the integral whole—the Universal Spirit, from which it emanates.[1]

The Significance of The Pilgrim and the Monad

The word "Pilgrim" further signifies a wayfarer (says the dictionary), although here it means also an individual entity pursuing a specific journey along the pathway of evolution and having as its goal the attainment of perfection in the human stage of evolution. Although a monad means an indestructible unit (since the word is derived from the Greek *monas,* a unit), it is often referred to as the two in one, because it signifies the union of man's two higher principles or components, namely Ātman, the divine spirit, and its upādhi (an emanational vehicle) Buddhi, the spiritual soul. "Soul" in this sense signifies the vehicle of the spirit. Specifically, the term Monad should be defined by two words, whether in English or in Sanskrit: "Eternal Pilgrim" or Ātma-Buddhi; or, if a symbolic term is desired, "a Spark of Eternity."

Continuing the exposition of the term "monad," *The Secret Doctrine* states:

> In reference to the Monads, the reader is asked to bear in mind that Eastern philosophy rejects the Western theological dogma of a newly-created soul for every baby born, as being as unphilosophical as it is impossible in the economy of Nature. There must be a limited number of Monads evolving and growing more and more perfect through their assimilation of many successive personalities, in every new Manvantara. This is absolutely necessary in view of the doctrines of Rebirth, Karma, and the gradual return of the human Monad to its source—absolute Deity.[2] Thus although the hosts

1. SD I:16 fn.; I:82, 6-vol. ed.; I:45, 3rd ed.

2. The return of the human Monad to its source is considered in Chapter 3.

> of more or less progressed Monads are almost incalculable, they are still finite, as is everything in this Universe of differentiation and finiteness.[3]

Then there is the "human monad." This term signifies the monad which uses the vesture of the human kingdom for an embodiment upon the earth—or, stated technically, the monad manifesting in that form of *prakriti* called the human kingdom. Prakriti here signifies the substance composing the vehicles or forms which manifest in the stage of evolution represented by the human kingdom.

Another term should be explained: manvantara. This means a period of activity, or a cycle of evolution of a planetary system. Literally, the Sanskrit word means "between Manus." A Manu signifies a regent of a planetary system. Furthermore, fourteen Manus partake of this regency. The total time-period of their regency, which includes the reigns and interregnums of the Manus, is stated to endure for a period of one thousand Mahāyugas. As a Mahāyuga consists of 4,320,000 years, the total time-period of the activity of a planetary system (or a Manvantara) is 4,320,000,000 years. This is also called a Day of Brahmā, or a planetary Kalpa. Following this period of activity there is a corresponding period of rest, which is designated a Pralaya, or a Night of Brahmā. Pralaya literally means a dissolution—a dissolution of forms. Its time period is equal to that of a Manvantara.

Thus *The Secret Doctrine* postulates that even though the hosts of Monads peopling the earth are almost incalculable, there is nevertheless a definite number of them. Moreover, the Monads mentioned in the quotation are the human hosts or the human kingdom; nevertheless it should be borne in mind that all the kingdoms of nature are composed of hosts of monads; yes, even the atoms:

> . . . Each atom is of course a *soul*, a monad, a little universe endowed with consciousness, hence with *memory*.[4]

---

3. SD I:171; I:224, 6-vol. ed.; I:194, 3rd ed.

4. Ibid. II:672; IV:241, 6-vol. ed.; II:709, 3rd ed.

> The Monads (Jīvas) are the Souls of the Atoms, both are the fabric in which the Chohans (Dhyānis, *gods*) clothe themselves when a form is needed.[5]

Therefore it may be stated that monads animate all living things, including man. Furthermore, we are provided with a clue to the origin of the monads, since their source is mentioned: Absolute Deity, or, again, the Universal Spirit—Paramātman in Sanskrit. Also postulated are the doctrines of Rebirth and Karma. The doctrine of Rebirth means that every monad which comes into incarnation on earth must have lived previously and consequently must have produced causes which have to be worked out in the sphere where they originate. This exemplifies the doctrine of Karma. Thus the monads come to the earth because they have previously lived on the earth, in order to take care of the "unfinished business" which they had left undone during former lives on the earth.

Since every man then returns to the physical world, which he inhabits from time to time in order to perfect his evolutionary status, is it not logical to propose the idea, by analogy, that the world upon which he lives is itself the result of a previous world's incomplete period of evolution? And, further, that the nature or condition of our present earth is the result of causes engendered by a corresponding previous world's existence? *The Secret Doctrine* asserts that this very thing did occur, that the earth is the progeny of a previous planet, and that this former planet, now dead, is the present earth's satellite, the moon. Hence the earth is sometimes called in esoteric literature, "the child of the moon."

> It thus becomes apparent how perfect is the analogy between the processes of Nature in the Kosmos[6] and in the individual man. The latter lives through his life-cycle, and dies. His "higher principles," corresponding in the development of a planetary chain to the cycling Monads, pass into Devachan, which corresponds to the

---

5. Ibid. I:619; II:344, 6-vol. ed.; I:679, 3rd ed.

6. Kosmos spelled with a "k" has reference to the Universe; *cosmos* refers to a solar system or a world.

> "Nirvāna" and states of rest intervening between two chains. The man's lower "principles" are disintegrated in time and are used by Nature again for the formation of new human principles, and the same process takes place in the disintegration and formation of Worlds. Analogy is thus the surest guide to the comprehension of the Occult teachings.[7]

An explanation of the terms used in the quotation may be useful. In the after-death states the monad is regarded as a trinity, signifying man's deathless higher principles, namely Ātma-Buddhi plus its upādhi (emanational vehicle) termed the Reincarnating Ego — which is technically referred to as Higher-Manas, or Buddhi-Manas. These three principles constitute man's immortal components; they are also designated the "higher triad." It is the portion here referred to as the Reincarnating Ego which experiences the state of Devachan—a Tibetan word meaning happiness. It signifies a condition of bliss in the after-death states during the period between incarnations on earth. Nirvāna means (briefly) a state of bliss superior to that of Devachan during the period of a planet's period of rest, which in Sanskrit is termed a pralaya. As pointed out in the quotation, the state of Devachan for the Reincarnating Ego corresponds to the state of Nirvāna for the monads. The latter state is described as "a state of rest between two chains." The two chains represent here two planetary systems, signifying the Earth planetary system and the Moon planetary system; the word chain refers to a series of seven globes. Every planetary system is composed of seven globes, in accordance with a postulate which is presented in *The Secret Doctrine*:

> Every sidereal body, every planet, whether visible or invisible, is credited with six companion globes.[8]

The six companion globes are represented as manifesting on superior planes, hence are regarded as "subjective globes" in comparison with our objective sphere or globe of physical

---

7. SD I:173; I:226, 6-vol. ed.; I:196, 3rd ed.

8. Ibid. I:158; I:213, 6-vol. ed.; I:182, 3rd ed.

substance.[9]

In regard to the reference to man's higher and lower principles, the Esoteric Philosophy regards man as composed of seven principles, which are referred to symbolically as the Three Fires—the immortal components—and the Four Flames—the mortal constituents. These may be shown as follows:

The Seven Principles

The Three Fires:	Ātman—The Divine Spirit
	Buddhi—The Spiritual Soul (also designated the Discriminating Principle)
	Manas—The Mind Principle
The Four Flames:	Kāma—The Energic Principle (more often referred to as the Desire Principle)
	Prāna—The Vital Principle (the Principle of Life)
	Linga-śarīra—The Model Vehicle (also termed the Etheric Double)
	Sthūla-śarīra—the Physical Body

The Three Fires have already been referred to as man's higher principles, namely the Monad (Ātma-Buddhi) and its emanational vehicle (or upādhi), the Reincarnating Ego which, in the after-death state, experiences the state of Devachan. With regard to the lower principles, it was stated that when death takes place the mortal constituents in time are disintegrated—signifying that each principle returns to the Source from which it was derived. Therefore, in due time the Kāma principle returns to the source from which it emanated—the Universal Śakti, or the Universal Will-Force or universal energy. Prāna reverts to Jīva—the universal Life-Principle. The Linga-śarīra goes back to its source, Ākāśa, the kosmical spirit-substance.[10] As for the physical body, since it is no longer vitalized by the principles

9. This will be explained in the next section in connection with the Three Classes of Monads.

10. Cf. ML, pp. 71-2.

just enumerated, it discontinues its up-building processes, because the life-atoms composing the Sthūla-śarīra (the physical body) return to the kingdoms from which they had been drawn.

As well as being called the four lower principles, these mortal constituents are usually referred to as the lower quaternary, and when functioning in connection with the lower mind are then designated the personality, in contradistinction to the individuality, which represents the higher mind or the Reincarnating Ego.

The Three Classes of Monads

The monads now inhabiting the earth were once hosts of monads existing on the moon. In fact *The Secret Doctrine* makes clear that the monads are divisible into three classes. Since the theme of the succeeding chapters is based upon the concepts brought forward in this initial chapter, an explanation regarding the three classes of monads is of prime importance.

> The Monadic Host may be roughly divided into three great classes:
>
> 1. The most developed Monads (the Lunar Gods or "Spirits," called, in India, the Pitris), whose function it is to pass in the first Round [of the Earth Chain] through the whole triple cycle of the mineral, vegetable, and animal kingdoms in their most ethereal, filmy, and rudimentary forms, in order to clothe themselves in, and assimilate, the nature of the newly formed chain. They are those who first reach the human form (if there can be any form in the realm of the almost subjective) on Globe A in the First Round. It is they, therefore, who lead and represent the human element during the second and third Rounds, and finally evolve their shadows at the beginning of the Fourth Round for the second class, or those who come behind them.[11]

The First Class of Monads: A clarification of the terms used in the quotation will assist in understanding the concepts

---

11. SD I:174; I:227, 6-vol. ed.; I:197, 3rd ed.

which are to be presented. "The most developed monads" represent the beings who were functioning in the human stage of evolution. In the esoteric philosophy the human stage of evolution represents the functioning of the mind principle, expressed technically thus:

> Between man and the animal—whose Monads (or Jīvas) are fundamentally identical—there is the impassable abyss of Mentality and Self-consciousness.[12]

The human stage of evolution is also designated the Human Kingdom, or the Human Life-Wave. "The most developed monads" were the humans who existed on the former world or planetary system which is now known as the moon. Those most developed monads are referred to as Lunar Gods because they had successfully accomplished the human cycle of evolution on the moon chain by attaining perfection in the human stage; they were therefore ready to graduate from the human kingdom at the conclusion of the life-cycle on that former planetary system before it passed into the moon-stage. These most developed monads are also called the Lunar Pitris. Pitri is a Sanskrit word meaning "father."

The next concept presented in the antecedent quotation had reference to the First Round. A Round signifies the completion of one cycle of evolution consisting of the passage of a life-wave, or a kingdom, through each one of the seven globes of a planetary system (such as the Earth planetary system) from the first to the seventh of the series. This circuit is made seven times.

> The evolution of life proceeds on these seven globes or bodies from the first to the seventh in Seven Rounds or Seven Cycles.[13]
>
> Our Earth, as the visible representative of its invisible superior fellow globes, its "lords" or "principles," has to live, as have the others, through seven Rounds.[14]

---

12. Ibid. II:81; III: 90, 6-vol. ed.; II:85, 3rd ed.

13. Ibid. I:159; I:213, 6-vol. ed.; I:182, 3rd ed.

14. Ibid. I:159; I:213, 6-vol. ed.; I:183, 3rd ed.

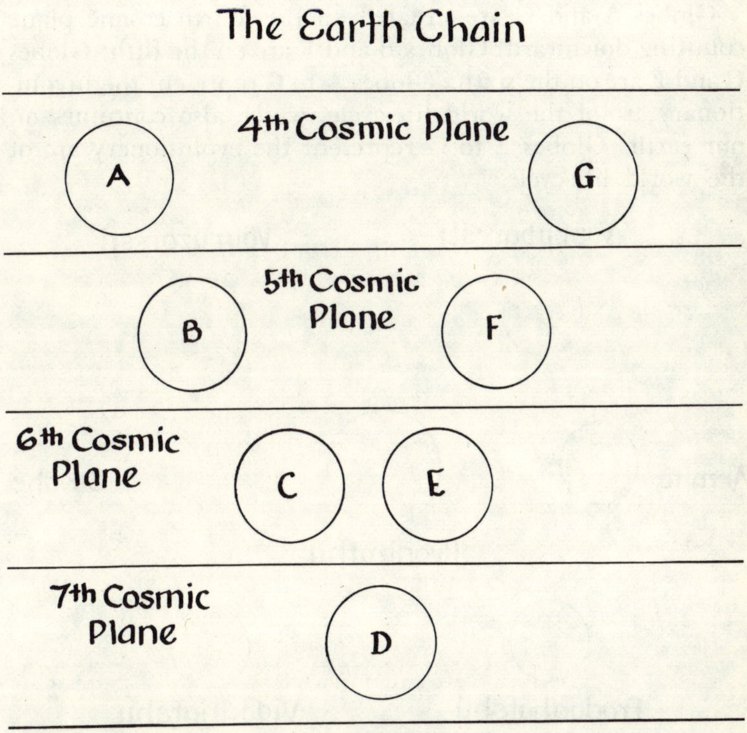

DIAGRAM 1

The seven globes of a planetary system are referred to as Globes A, B, C, D, E, F, G. Our Earth is Globe D, the fourth in the series. Its "superior fellow globes" are invisible to us because they are situated in subjective realms, regarded in the esoteric philosophy as divisible into cosmic planes not perceivable by our ordinary senses. Globe D, our earth, is situated on the lowest or seventh cosmic plane, the physical. Our senses function solely on this plane. In Occultism a plane is defined as

> . . . the range or extent of some state of consciousness, or the state of matter corresponding to the perceptive

powers of a particular set of senses or the action of a particular force.[15]

Globes A and G are situated on the fourth cosmic plane counting downward; Globes B and F are on the fifth; Globes C and E are on the sixth. Globes A to C represent the involutionary arc of the world life-cycle, which also continues on our earth. Globes E to G represent the evolutionary arc of the world life-cycle.

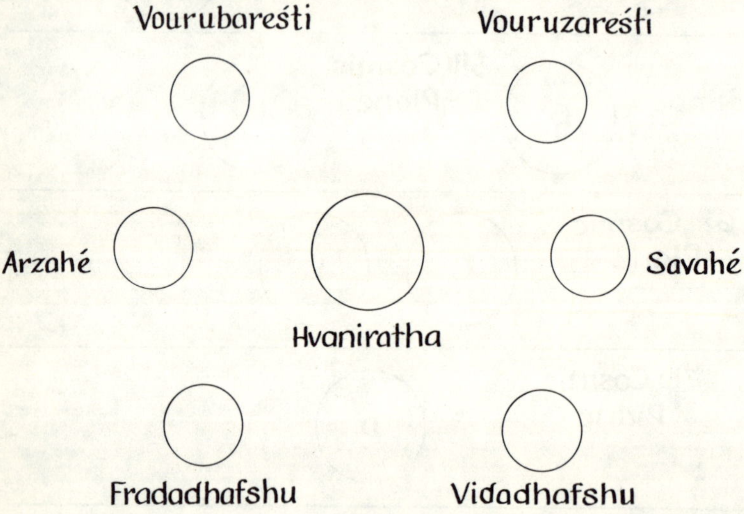

DIAGRAM 2

In the ancient Persian work known as the *Vendīdād* the seven globes of the Earth Chain were named Vourubareśti, Arzahé, Fradadhafshu, Hvaniratha (our Earth, Globe D), Vidadhafshu, Savahé, Vouruzareśti. The globes are situated on Karshvars, a term equivalent to cosmic planes: two of them, Vourubareśti and Vouruzareśti, symbolically lie in the north; two, Vidadhafshu and Fradadhafshu, in the south; Savahé and Arzahé in the east and west respectively.[16]

15. *The Key to Theosophy,* p. 360. (Subsequent references to this work will be designated *Key.*)

16. SD II:759; IV:328, 6-vol. ed.; II:802, 3rd ed.

We come again to a consideration of the first class of monads, defined as "the most developed, whose function it is to pass in the first Round through the whole triple cycle." "The triple cycle" refers to the evolutionary stages undergone in each of the three kingdoms of nature which were specified: the mineral kingdom, the plant or vegetable kingdom, the animal kingdom. In this enumeration the human kingdom represents the fourth evolutionary cycle of the monads. Each one of the three kingdoms undergoes seven evolutionary stages on each one of the seven globes of the planetary chain beginning on Globe A and concluding on Globe G. The passage of the monads from Globe A to Globe D is termed the Descending Arc of the circuit. The passage through Globes E to G is designated the Ascending Arc of the evolutionary cycle. However, the first class of monads, instead of undergoing evolutionary stages in the "triple cycle" rapidly recapitulate stages of development already attained in previous manvantaras. Therefore, this rapid passage should not be regarded as comparable to an evolutionary stage of development, inasmuch as the rūpas of all the kingdoms in their first elementary states in the first Round were "most ethereal, filmy, and rudimentary forms." That is to say, their forms (rūpas) were non-material as we normally regard matter. "The newly formed chain" signifies the seven globes of the earth planetary system.

It should be borne in mind that the humans during the first Round had a totally different rūpa (or form) from the bodies of the present cycle. Witness this passage from *The Secret Doctrine*:

> In truth, the "Man" on Globe A during the First Round is no Man, but only his prototype or dimensionless image from the astral regions.[17]

Then, too, it is the most developed monads who lead the human lifewave during the Second and Third Rounds of the evolutionary circuit of the seven globes. Continuing their leadership, in the present Fourth Round they "evolved their

---

17. Ibid. I:175, fn.; I:228, 6-vol. ed.; I:199, 3rd ed.

shadows," signifying that they produced the "pattern forms" which ultimately resulted in the present human bodies, as will be related in succeeding chapters.

The Second Class of Monads: We now come to a consideration of the second class of monads, which deals with the subject of monadic evolution. Attention is directed to the fact that in this exposition reference is *not* made to the physical bodies of human beings, whether in this present stage of evolutionary development or in the early Rounds, which is the subject presently under consideration. That such a distinction is required is observable in this passage:

> The evolution of the *external* form or body round the *astral* is produced by the terrestrial forces, just as in the case of the lower kingdoms; but the evolution of the internal or real MAN is purely spiritual. It is now no more a passage of the impersonal Monad through many and various forms of matter—endowed at best with instinct and consciousness on quite a different plane—as in the case of external evolution, but a journey of the "pilgrim-soul" through various *states* of *not only matter* but Self-consciousness and self-perception, or of *perception* from apperception.[18]

Apperception is a term used by Leibnitz intending to convey the meaning that "nerve-sensation expresses the state of monadic consciousness through all the Kingdoms up to Man."[19]

The closing sentence of the definition covering the first class of monads referred to the Lunar Pitris as evolving "their shadows at the beginning of the Fourth Round for the second class" of monads. The explanation is now given regarding the state and condition of the "shadows" which were provided by the Lunar Pitris: they were astral rūpas (forms). In the course of ages, "physical coverings" were "produced by the terrestrial forces" culminating in an "external form or body round the astral." This process was

18. Ibid. I:175; I:228, 6-vol. ed.; I:198, 3rd ed.

19. Ibid. I:179; I:231, 6-vol. ed.; I:201, 3rd ed.

achieved during the Third Root-Race, about ten million years ago.

The second class of monads is defined in this manner:

> Those Monads that are the first to reach the human stage during the three and a half Rounds, and to become men.[20]

The monads in this category represent two grades or stages of evolution: (a) those monads on the Moon-Chain who had *not* attained the stage of evolution that enabled them to graduate from the human kingdom at the conclusion of the seven-round life-cycle of the lunar planetary system, and (b) those monads on the Moon-Chain who had attained sufficient evolutionary development in the animal kingdom to enable them to graduate from that kingdom and so to reach the human stage on our earth. Both these classes of monads were able to enter the human kingdom during the first three-and-a-half Rounds of the Earth Chain.

The significance of the three-and-a-half Rounds should not be overlooked. It represents the middle point of the manvantara, or the half-way mark in the seven-round life-cycle of the planetary system. From the standpoint of the time period, it occurred on this earth during the period of the Atlantean culture. Expressed technically, the middle point of the Circle of Necessity of the Seven-Round Cycle was reached on this fourth globe of the earth planetary system, on the fourth Continent, during the fourth subrace of the Fourth Root-Race.

The Third Class of Monads: In this present classification of the monads, a different concept is presented in connection with the third class: a specific period is stated in connection with a definite evolutionary phase. Although the Esoteric Philosophy asserts that motion is eternal and, consequently, that there is no possibility of anything remaining permanently in a state of quiescence, there is a definite time period in which progress may be made. When this time period has elapsed evolutionary progress of a certain type is halted,

20. Ibid. I:174; I:227, 6-vol. ed.; I:197, 3rd ed.

namely availability of entrance into the human kingdom. This is instanced in connection with the third class of monads, who are termed:

> The laggards: the Monads which are retarded, and which will not reach, by reason of Karmic impediments, the human stage at all during this cycle or Round.[21]

The present cycle is the Fourth Round. The "Karmic impediments" are applicable to certain monads simply because those monads have not attained sufficient individualized evolutionary development to maintain progress in the human kingdom during the evolutionary Arc of Ascent.

It has sometimes been asked whether any members of the present human kingdom are included in this third class of monads. Such a question could not have arisen had the passage conveying the recapitulation of the three monadic classes been understood. *The Secret Doctrine* clearly gives the reason why this third class cannot be included in the human kingdom during this manvantara; also why the term "laggards" is applied to this class of monads.

Let us recapitulate.

> The most developed Monads (the lunar) reach the human germ-stage in the first Round: become terrestrial, though very ethereal human beings towards the end of the Third Round, remaining on it (the globe) through the "obscuration" period [see below] as the seed for future mankind in the Fourth Round, and thus become the pioneers of Humanity at the beginning of this, the Fourth Round. Others reach the Human stage only during later Rounds, i.e., in the second, third, or first half of the Fourth Round. And finally the most retarded of all, i.e., those still occupying animal forms after the middle turning-point of the Fourth Round—will not become men at all during this Manvantara. They will reach to the verge of humanity only at the close of the seventh Round to be, in their turn, ushered into a new chain after pralaya—by older pioneers, the progenitors of humanity, or the Seed-Humanity (Sishta), viz., the

21. Ibid. I:175; I:227, 6-vol. ed.; I:198, 3rd ed.

men who will be at the head of all at the end of these Rounds.[22]

The "obscuration period" is best explained by outlining the stages of evolutionary development of the human kingdom which take place during the Fourth Round. The Fourth Round begins when the monads comprising the human life-wave enter Globe A of the Earth Chain or planetary system. They undergo seven major stages of evolutionary development on that globe. When the seventh major evolutionary stage has been accomplished there, the hosts of human monads leave Globe A and enter Globe B. When all the hosts of human monads have left Globe A—except for one special class called called Śishtas, or the "divine mortals," or in the case of the human kingdom the "Seed-Humanity," signifying "the seed for future mankind"—the vacated globe becomes dormant. This has been termed an "obscuration period." The monads undergoing evolutionary stages in all the kingdoms below the human kingdom have also left Globe A, except for the appropriate Śishtas pertaining to each kingdom. Śishta is a Sanskrit word literally meaning "remainders." The Śishtas represent the highest representatives of each class or kingdom. They remain behind in order to provide forms for the incoming monads on the next cycle of evolutionary activity (this is termed a Round). The Śishtas do not undergo any evolutionary development during the period of dormancy; that is to say no new phyla (to use a biological term meaning racial stocks) will be produced from the seeds of life remaining on the vacated globe during that time-period.

After undergoing seven major developmental stages on Globe B, the hosts of human monads leave Globe B, again except for the Śishtas, and enter Globe C. The vacated globe, as in the case of Globe A, also becomes dormant. After fulfilling seven major developmental stages of evolution on Globe C, the hosts of human monads leave Globe C, except for the Śishtas, and enter Globe D. Globe C then becomes quiescent.

The Śishtas are not the second class of monads. The latter

22. Ibid. I:182; I:233-4, 6-vol. ed.; I:204-5, 3rd ed.

are those, as said, who "reach the human stage during later Rounds, that is, in the second, third, or first half of the Fourth Round." As for the third class of monads, they will not enter the human kingdom during this Manvantara (the seven-round life-cycle) for the "the door is closed for this Cycle and the balance struck."

The Secret Doctrine stresses the importance of the fourth globe in the series of the seven—which is our earth—because it occupies a unique position. It is situated on the seventh and lowest cosmic plane.

> Unlike the others, the Fourth has no "sister" Globe on the same plane as itself, and it thus forms the fulcrum of the "balance" represented by the whole chain. It is the sphere of final evolutionary adjustments, the world of Karmic scales, the Hall of Justice, where the balance is struck which determines the future course of the Monad during the remainder of its incarnations in the cycle. And therefore it is, that, after this central turning-point has been passed in the Great Cycle,—i.e., after the middle point of the Fourth Race in the Fourth Round on our Globe—no more Monads can enter the human kingdom. The door is closed for this Cycle and the balance struck. For were it otherwise—had there been a new soul created for each of the countless milliards of human beings that have passed away, and had there been no reincarnation—it would become difficult indeed to provide room for the disembodied "Spirits"; nor could the origin and cause of suffering ever be accounted for.[23]

Attention is especially directed to the second sentence of the above quoted passage. It is apt to be overlooked. It calls attention to the fact that here, on *this* earth, the *final* evolutionary adjustments are to be made. Then, too, the closing portion of the sentence is pregnant with meaning, for it is on this earth "where the balance is struck which determines the future course of the Monad during the remainder of its incarnations in the cycle." Then, in the third sentence the declaration is made: ". . . no more Monads can enter the human kingdom" for *this* manvantara.

---

23. Ibid. I:182-3; I:234, 6-vol. ed.; I:205, 3rd ed.

The Descending and Ascending Arcs of Evolution

During the course of the seven-round circuit or life-cycle of a planetary system, the central turning point of the Great Cycle is reached. This marks the lowest point in the arc of descent. It is reached during the mid-point of the fourth major evolutionary phase of the human race. This phase is termed the fourth subrace of the Fourth Root-Race on the Fourth Round of the fourth globe of a planetary system, which is our earth, Globe D. During the course of the arc of descent the involutionary process of spirit is taking place: spirit is progressively immersed in matter. On the other hand the Arc of Descent, the Descending Arc or the Shadowy Arc, represents the process of evolution of matter. When the culminating point in the Descending Arc is reached the reverse process begins: an evolution of spirit and a concurrent involution of matter begins to take place. That is to say a spiritualizing evolutionary process of the rūpas (or forms), which the monads assume, gradually eventuates. This process in connection with the evolution of spirit, as indicated earlier, is called the Arc of Ascent, the Ascending Arc or the Luminous Arc, during which spirit is progressively liberated from form with expanding self-consciousness. The process will continue, gathering greater and greater momentum during each Round, until the final culmination at the end of the Seventh Round comes to pass. Then the involution of matter, as well as the evolution of spirit, will be completed.

The process may be illustrated by an example, represented in an ancient symbol portraying the significance of the descending and ascending arcs. Strangely enough, this symbol, in a somewhat modified form, is in use nowadays by the medical profession and is also used in the badge of the army medical corps. In ancient times it was used as the herald's staff. The Greek word for the staff, *karykion,* is rendered in Latin as caduceus. Hermes, and later Mercury, in Greece and Rome, was represented as wearing winged sandals and bearing the caduceus. However, the origin and meaning of the

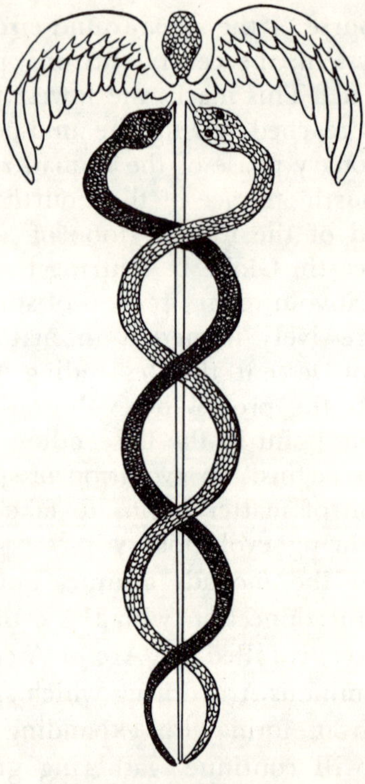

DIAGRAM 3

caduceus are well described in a Commentary from the Book of Dzyan:

> The trunk of the Aśvattha (the tree of Life and Being, the Rod of the caduceus) grows from and descends at every Beginning (every new manvantara) from the two dark wings of the Swan (Hansa) of Life. The two Serpents, the ever-living and its illusion (Spirit and matter) whose two heads grow from the one head between the wings, descend along the trunk, interlaced in close embrace. The two tails join on earth (the manifested Universe) into one and this is the great illusion, O Lanoo![24]

24. Ibid. I:549; II:273, 6-vol. ed.; I:600, 3rd ed.

In the Scandinavian cosmogony, Yggdrasil (the World Ash-tree), like the Aśvattha, stands for a manvantara, inasmuch as it flourishes throughout a period of activity. The two serpents entwined around the Aśvattha (originally), now replaced by a rod, represent the Descending and Ascending Arcs in the Great Cycle of Life and Being during a manvantara. The dark serpent on the left symbolizes the Arc of Descent, the light serpent on the right, the Arc of Ascent.

From the standpoint of materiality, the grossest point both in the composition of the human physical body itself and in the power and strength of the vehicle, was reached at the lowest point in the Arc of Descent, namely during the Fourth Round in the fourth subrace of the Fourth Root-Race, or the fourth major evolutionary phase of humanity. As the host of human monads are now using vehicles representing the fifth major evolutionary phase (termed the Fifth Root-Race), the physical body is no longer as gross or as powerful as it was during the Fourth Root-Race of the Atlantean era. Eventually the evolution of spirit will predominate as the involution of matter proceeds apace.

> On the descending arc it is the spiritual which is gradually transformed into the material. On the middle line of the base, Spirit and Matter are equilibrized in Man. On the ascending arc Spirit is slowly reasserting itself at the expense of the physical, or matter, so that at the close of the seventh Race of the Seventh Round, the Monad will find itself as free from matter and all its qualities as it was in the beginning; having gained in addition the experience and wisdom, the fruition of all its personal lives, without their evil and temptation.[25]

25. Ibid. II:180-1; III:187, 6-vol. ed.; II:190-1, 3rd ed.

2.

THE SOURCE OF THE IMMEMORIAL WISDOM

IN CONSIDERING the source of the wisdom from which present-day Theosophy springs, one soon discovers that many scriptures are mentioned in *The Secret Doctrine*. Each one has its own method of presenting ideas. H. P. Blavatsky used these ancient scriptures as a means of showing that there is a golden thread which unites them all, thus indicating that they have come from a single source. This enables the student to recognize that there is a fundamental truth which each system of thought seeks to express in one way or another. In thus writing on the occult lore presented in these scriptures, as well as about the mystery schools which had flourished in various parts of the world, H. P. Blavatsky was of course aware of the fact that there was a system of philosophy, often referred to as the esoteric doctrine, which had been formulated so as to convey this immemorial wisdom to those who were entitled to receive it. Thus we find her writing on this very theme:

> . . . there is reason to call the trans-Himalayan esoteric doctrine Chaldeo-Tibetan. And, when we remember that the *Vedas* came—agreeably to all traditions—from the Manasarovara Lake in Tibet, and the Brahmins themselves from the far North, we are justified in looking on the esoteric doctrines of every people who once had or still have it—as having proceeded from one and

the same source; and to thus call it the "Aryan-Chaldeo-Tibetan" doctrine, or Universal Wisdom-Religion.[1]

The Vedas and the Upanishads

Then are the Vedas to be regarded as the source of this Universal Wisdom-Religion? Not as being the source of this wisdom, but as the means whereby the immemorial wisdom was presented to the early representatives of the Aryan Race, before its members came into India from the far north. The quotation continues:

> The *Vedas,* Brahmanism, and along with these, Sanskrit, were importations into what we now regard as India. They were never indigenous to its soil. There was a time when the ancient nations of the West included under the generic name of India many of the countries of Asia now classified under other names.[2]

Very little is known concerning how the Vedas came to be written. The Brahmans ascribe their preparation to Vyāsa, for he is regarded as a compiler, not as the author of the Vedas. There is no doubt that they are the most ancient works of the Hindus. They are mentioned in the Laws of Manu (the *Mānava-dharma-śāstra*), itself an old Sanskrit scripture. In this work only three of the Vedas are mentioned. Orientalists conclude that the fourth Veda, known as the *Atharva-Veda,* is a much later production than the first three Vedas—the *Rig-Veda,* the *Yajur-Veda,* and the *Sāma-Veda.* These three are regarded in the Laws of Manu as *sanātanaṃ trayam brahma*—"the eternal triple divine knowledge." The word *veda* itself signifies knowledge, sacred knowledge; it is derived from the verbal root *vid,* to know.

---

1. "The Aryan-Arhat Esoteric Tenets on the Sevenfold Principle in Man" by T. Subba Row; Editorial appendix by H. P. Blavatsky: *The Theosophist,* January, 1882; *H. P. Blavatsky Collected Writings,* III:419. (Subsequent references to the latter work will be designated BCW.) (The term *Aryan* in this text denotes that branch of the Indo-European peoples which spread over northern India at the beginning of the Vedic Age. It is not to be confused with the term as used in Germany during World War II.)

2. BCW, III:420.

The oldest of the three Vedas is the *Rig-Veda*—literally the Hymn-Veda, or the Veda of praise. It is a collection of 1017 hymns, arranged in ten mandalas—divisions or "books"—written in meter, in the most ancient form of Sanskrit and intended to be recited aloud. The *Yajur-Veda* is the sacrificial Veda (*yajus* means a sacrifice): it is a collection of mantras which are to be recited during the observance of sacrifices. The *Sāma-Veda* is the Veda of chants (*sāman* signifies chant), to be used in connection with religious rites. The later *Atharva-Veda* consists principally of formulas and spells to be used in order to counteract diseases and calamities. The word *Atharva* signifies a priest who uses fire and soma (the sacred juice of the soma plant) in his religious observances.

Subba Rao, an able exponent of Brahmanical philosophy, stated:

> The *Vedas* were perhaps compiled mainly for the use of the priests assisting at public ceremonies, but the grandest conclusions of our real secret doctrine are therein mentioned. I am informed by persons competent to judge of the matter, that the *Vedas* have a distinct dual meaning—one expressed by the literal sense of the words, the other *indicated by the meter and the svara* [intonation] which are, as it were, the life of the Vedas.[3]

An important consideration in connection with the Vedas is that each of the four scriptures is composed of two parts: (1) the Mantra—words of prayer and adoration; (2) the Brāhmana—directions for the observance of the ceremonies during which the mantras were to be recited and explanation of the meaning of the legends associated with the mantras. Moreover, concerning the Brāhmana portion of the Vedas, further exposition of the legends was given in the *Āranyakas* and *Upanishads;* consequently these works are regarded as pertaining to Vedic literature. The literal meaning of both these words indicates the nature of the scriptures. *Āranyaka* signifies forest-born (*āranya,* a forest), hence writings which came into being in a forest, where the Vedas were

---

3. Ibid. III:401.

studied. The term *Upanishad* consists of a prepositional prefix, *upa,* meaning near; *ni,* down; *shad,* from the verbal root *sad,* to sit. Hence the term may be rendered "what is received when seated near," thus conveying the idea that instruction was received when sitting near a teacher. At all events the *Upanishads* represent the exposition of the Vedas. With regard to the *Upanishads* H. P. Blavatsky wrote:

> Translated as "esoteric doctrine," or interpretation of the *Vedas* by the *Vedānta* methods. The third division of the *Vedas* appended to the *Brāhmanas* and regarded as a portion of *śruti* or "revealed" word. They are, however, as records, far older than the *Brāhmanas.* . . . It is from these treatises of the *Upanishads*—themselves the echo of the primeval Wisdom-Religion—that the Vedānta system of philosophy has been developed. . . . The accepted number of these treatises is 150, though now no more than about twenty are left unadulterated. They treat of very abstruse, metaphysical questions, such as the origin of the Universe; the nature and the essence of the Unmanifested Deity and the manifested gods; the connection, primal and ultimate, of spirit and matter; the universality of mind and the nature of the human Soul and Ego. . . . There was a time, aeons before the Brahmans became a caste, or even the *Upanishads* were written, when there was on earth but one "lip," one religion and one science, namely, the speech of the gods, the Wisdom-Religion and Truth.[4]
>
> Not only are the *Upanishads* a secret doctrine, but in dozens of other works as, for instance, in the *Aitareya Āranyaka,* it is plainly expressed that they contain *secret doctrines,* that are not to be imparted to any one but a *Dvija* Brahman.[5]

---

4. *Theosophical Glossary* pp. 353-4. (Subsequent references to this work will be designated TG.) The Vedānta is considered to be the flowering or greatest development of the Uttara-Mīmānsā Darśana—one of the six Hindu systems of philosophy (termed darśanas). Vedānta signifies "the complete knowledge of the Veda"; *uttara* means higher, excellent, predominant; mīmānsā means profound thought, examination (of Vedic thought). This darśana is also referred to as Brahma-jñāna—"spiritual knowledge of Brahman." Śankarāchārya taught the Advaita system of Vedantic philosophy.

5. "Was Writing Known Before Pānini?" by a Chela, *The Theosophist,* October 1883, BCW V:297.

Along with these Hindu scriptures, mention should be made of the *Purānas,* regarded as containing legendary stories of ancient India. In fact the *Purānas* are a collection of eighteen Brahmanical writings, allegorical and symbolical in character. They treat of the creation, passing, and renovation of worlds, as well as the genealogy and activities of deities and sages and heroes. The word *purāna* signifies "ancient"; the writings are composed in epic couplets. In addition to the eighteen principal volumes there are eighteen subordinate works known as *upa-purānas.* Although written in the form of fables there is no doubt that there is a great deal of esotericism in the ancient writings. H. P. Blavatsky comments:

> To comprehend the esoteric meaning of ancient Brahmanical literature, one has to be in possession of the key to the Brahmanical Code. To master the conventional terms used in the *Purānas,* the *Āranyakas* and *Upanishads* is a science in itself, and one far more difficult than even the study of the 3,996 aphoristical rules of Pānini, or his algebraical symbols.[6]
>
> The . . . teachings of the SECRET DOCTRINE, supplemented by universal traditions, must now have demonstrated that the *Brāhmanas* and *Purānas,* the Gāthās and other Mazdean Scriptures, down to the Egyptian, Greek, and Roman, and finally to the Jewish Sacred records, all have the same origin. None are meaningless and baseless stories, invented to entrap the unwary profane: all are allegories intended to convey, under a more or less fantastic veil, the great truths gathered in the same field of prehistoric tradition. . . . It has been shown, on the evidence of the whole ancient literary world, and the intuitional speculations of more than one philosopher and scientist of the later ages, that the tenets of our Esoteric Doctrine are corroborated by inferential as well as by direct proof in almost every case."[7]

---

6. "Historical Difficulty"—Why? *The Theosophist,* October 1883, BCW V:208.

7. SD II:409-10; III:408, 6-vol. ed.; II:428, 3rd ed.

Ancient Scriptures

However, further back in time than any of the scriptures which have come down to our day, there exists "a literary relic" according to H. P. Blavatsky. She first mentioned it on the opening page of her initial work, *Isis Unveiled*:

> There exists somewhere in this wide world an old Book—so very old that our modern antiquarians might ponder over its pages an indefinite time, and still not quite agree as to the nature of the fabric upon which it is written. It is the only original copy now in existence. The most ancient Hebrew document on occult learning—the *Siphra Dzeniouta*[8]—was compiled from it, and that at a time when the former was already considered in the light of a literary relic.[9]

Later, in *The Secret Doctrine* H. P. Blavatsky wrote regarding this archaic manuscript:

> The "very old Book" is the original work from which the many volumes of *Kiu-ti* were compiled. Not only this latter and the *Siphra Dzeniouta* but even the *Sepher Jezirah,* the work attributed by the Hebrew Kabalists to their Patriarch Abraham (!), the book of *Shu-king,* China's primitive Bible, the sacred volumes of the Egyptian Thoth-Hermes, the Purānas in India, and the Chaldean *Book of Numbers* and the *Pentateuch* itself, are all derived from that one small parent volume. Tradition says that it was taken down in *Senzar,* the secret sacerdotal tongue, from the words of the Divine Beings, who dictated it to the sons of Light, in Central Asia, at the very beginning of the 5th (our) race; for there was a time when its language (the *Sen-zar*) was known to the Initiates of every nation, when the forefathers of the

---

8. H. P. Blavatsky refers to this work (also spelled *Siphra Dtzeniuthah*) as "the oldest book of the Kabbalists." It has been rendered into Latin by C. K. von Rosenroth under the name of *Liber Mysterii* in his *Kabbala denudata.* An English translation is available in M. Mathers's *Kabbalah Unveiled* under the section entitled "The Book of the Concealed Mystery," which again is one of the divisions of the *Zohar* ("Book of Splendor").

9. *Isis Unveiled,* I:1. (Subsequent references to this work will be designated *Isis.*)

> Toltec understood it as easily as the inhabitants of the lost Atlantis, who inherited it, in their turn, from the sages of the 3rd Race, the *Manushis,* who learnt it direct from the *Devas* of the 2nd and 1st Races.[10]

An elucidation of the ancient works quoted in this extract from *The Secret Doctrine* produces some interesting side-lights. (The *Kiu-ti* will be considered later.)

The *Sepher Jezirah* (also spelled *Yetzīrah*) signifies the "Book of Creation." It is another Kabalistic work, attributed to Rabbi Aqībah (who died 135 A.D.). H. P. Blavatsky described it as:

> . . . the most occult of all the Kabalistic works now in the possession of modern mystics.[11] Its alleged origin, of having been written by Abraham, is of course non-sense; but its intrinsic value is great. It is composed of six *Perakim* (chapters), sub-divided into thirty-three short *Mishnas* or Sections; and treats of the evolution of the Universe on a system of correspondences and numbers. Deity is said therein to have formed ("created") the Universe by means of numbers "by thirty-two paths (or ways) of secret wisdom," these ways being made to correspond with the twenty-two letters of the Hebrew alphabet and the ten fundamental numbers. These ten are the primordial numbers whence proceeded the whole Universe, and these are followed by the twenty-two letters divided into *Three Mothers,* the seven double consonants and the twelve simple consonants.[12]

The three mother letters are A, M, S; the seven double consonants signify the seven sacred planets; the twelve simple consonants stand for the twelve signs of the zodiac.

The *Shu-King* is described as the Book of Historical Documents, a collection of records regarded as antique in the time of Confucius (550-478 B.C.). In its preface, one hun-

---

10. SD I:xliii; I:64, 6-vol. ed.; I:26, 3rd ed.

11. "The *Sepher Yetzīrah* now known is but a portion of the original one incorporated in the Chaldean *Book of Numbers.*" SD III:199; V:205, 6-vol.ed.

12. TG, 165. The three mother letters stand for the three Elements: Air, Water, Fire.

dred books are mentioned, of which only fifty-nine are extant, the oldest going back to the twenty-third century B.C., the latest to the eighth. It has an astronomical interest also, for among other items there is a reference to a partial eclipse of the sun.

With regard to the sacred volumes of Thoth-Hermes, H.P. Blavatsky, first giving the meaning of the word Hermetic, wrote:

> Any doctrine or writing connected with the esoteric teachings of Hermes, who, whether as the Egyptian Thoth or the Greek Hermes, was the God of Wisdom with the Ancients, and, according to Plato, "discovered numbers, geometry, astronomy and letters."[13]
>
> The forty-two Sacred Books of the Egyptians, mentioned by Clement of Alexandria as having existed in his time, were but a portion of the Books of Hermes. (*Stromata,* Vol. II, p. 324) Iamblichus, on the authority of the Egyptian priest Abammon, attributes 1,200 of such Books to Hermes, and Manetho, 36,000.[14]
>
> There is also the additional evidence of Clemens Alexandrinus, that ought to be credited to some extent. Clemens testified to the existence of an additional 30,000 volumes of the Books of Thoth, placed in the library of the Tomb of Osymandias, over the entrance of which were inscribed the words, "A Cure for the Soul."[15]

Reference is made to the Hermetic books in *The Mahatma Letters to A. P. Sinnett* under Hermetic Philosophy in this manner:

> Hermetic Philosophy suits every creed and philosophy and clashes with none. It is the boundless ocean of Truth, the central point whither flows and wherein meet every river, as every stream—whether its source be in the East, West, North, or South. As the course of the river depends upon the nature of its basin, so the channel for communication of Knowledge must con-

---

13. Ibid. p. 140.

14. SD III:37; V:58, 6-vol. ed.

15. Ibid.

> form itself to surrounding circumstances. The Egyptian Hierophant, the Chaldean Mage, the Arhat, and the Rishi, were bound in days of yore on the same voyage of discovery and ultimately arrived at the same goal though by different tracks. . . . TRUTH has no ear-mark and does not suffer from the name under which it is promulgated—if the said object is attained.[16]

Another book of wisdom, given the name of the *Book of Numbers,* originated in Chaldea. This book should not be confused with the fourth book of Moses called "Numbers" in the Pentateuch (the first five books of the Old Testament). The Chaldean *Book of Numbers* is

> A work which contains all that is found in the *Zohar* of Simeon Ben-Jochai, and much more. It must be the older by many centuries, and in one sense its [the Zohar's] original, as it contains all the fundamental principles taught in the Jewish Kabbalistic works, but none of their blinds. It is very rare indeed, there being perhaps only two or three copies extant, and these in private hands.[17]
>
> We are not aware that a copy of this ancient work is embraced in the catalogue of any European library; but it is one of the "Books of Hermes," and it is referred to and quotations are made from it in the works of a number of ancient and medieval philosophical authors.[18]
>
> The *Books of Hermes* are the oldest repositories of numerical Symbology in Western Occultism.[19]
>
> In a word, the real, genuine Kabalah, the only original copy of which is contained in the Chaldean *Book of Numbers,* pertains to, and teaches about, the realm of spirit, not that of matter.[20]
>
> The use of all the seven keys to unlock the mysteries of Being in this life, and the lives to come, as in those

---

16. ML (Letter No. 85); p. 399; p. 393, 3rd ed.

17. TG, p. 75.

18. SD III:99, V:114, 6-vol. ed.

19. Ibid. III:100; V:115, 6-vol. ed.

20. "The Kabalah and the Kabalists," *Lucifer,* May 1892; BCW VII: 267.

which have gone by, shows that the Chaldean *Book of Numbers* and the *Upanishads* undeniably conceal the most divine philosophy—as it is that of the Universal Wisdom Religion.[21]

On Sanskrit and Devanāgarī

Orientalists are uncertain about the origin of Sanskrit. Some scholars designate a locality in which the early Aryans (who supposedly spoke Sanskrit) had their culture. Professor M. Monier-Williams in his celebrated dictionary states: "the Aryan languages (of which Sanskrit is the eldest sister) proceeded from a common but nameless and unknown parent . . . the locality may conjecturally be placed somewhere in the region of Bactria (Balkh) and Sogiana." (p. xii)

A Hindu chela prepared an article for *The Theosophist* entitled "Was Writing Known Before Panini?" in which he discussed the origin of Sanskrit, first prefacing his remarks by referring to the origin of languages:

> It can be proved that before the final division and classification of languages, there existed two languages in every nation: (*a*) the profane or popular language of the masses; (*b*) the sacerdotal or secret language of the initiates of the temples and mysteries—the *latter being one and universal.* Or, in other words, every great people had, like the Egyptians, its Demotic and its Hieratic writing and language, which had resulted first in a pictorial writing or the hieroglyphics, and later on in a phonetic alphabet. . . . The direct progenitor of the Vedic Sanskrit was the sacerdotal language (which has a distinct name among the initiates). The Vāch—its *alter ego* or the "mystic self," the sacerdotal speech of the initiated Brahman—became in time the mystery language of the inner temple, studied by the initiates of Egypt and Chaldea; of the Phoenicians and the Etruscans; of the Pelasgi and Palanquans; in short, of the whole globe. The appellation Devanāgarī is the synonym of, and identical with, the Hermetic *Neter-Khari*

21. Ibid. 260.

(divine speech) of the Egyptians.[22]

The writer of the article goes on to consider the subject of a language which came to be perfected or polished. This is the literal meaning of the word Sanskrit, which is indeed considered to be a perfected language because of its highly inflected and precise grammatical construction. He states that the Pelasgi were Aryans and then refers to Herodotus, who stated that the Pelasgi were the forefathers of the Greeks and that they spoke a "barbarous language." But all Greeks referred to any language other than Greek as "barbarous." The Hindu writer states that the "language mentioned by Herodotus was simply 'the primitive and now extinct Aryan tongue' that preceded the Vedic Sanskrit."[23] That is to say the language regarded as "the root of Sanskrit."

Devanāgarī also is considered by the writer. The word literally means "divine city writing," although its present meaning is rendered: "the written form of Sanskrit," that is to say, the alphabetical writing of Sanskrit:

> Real Devanāgarī—non-phonetic characters—meant formerly the outward symbols, as to say, *the signs used in the inter-communication between gods and initiated mortals.* Hence their great sacredness and the silence maintained throughout the Vedic and the Brahmanical periods about any object concerned with, or referring to, reading and writing. It was *the* language of the gods. If our Western critics can only understand what the Ancient Hindu writers meant by Rhutaliai, so often mentioned in their mystical writings, they will be in a position to ascertain the source from which the Hindus first derived their knowledge of writing.
>
> A secret language common to all schools of occult science once prevailed throughout the world.[24]

Continuing the subject of the Devanāgarī characters, H. P. Blavatsky further wrote concerning the sound value of the

22. *The Theosophist,* October 1883; BCW V:294, 297.

23. Ibid. V:301.

24. Ibid. V:306.

Sanskrit alphabet and its origin, as well as on the origin of the alphabets of other ancient languages:

> The Devanāgarī characters in which Sanskrit is generally written, have all that the Hermetic, Chaldean and Hebrew alphabets have, and in addition the Occult significance of the "eternal sound," and the meaning given to every letter in its relation to spiritual as well as terrestrial things. As there are only twenty-two letters in the Hebrew alphabet and ten fundamental numbers, while in the Devanāgarī there are thirty-five consonants and sixteen vowels, making altogether fifty-one simple letters, with numberless combinations in addition, the margin for speculation and knowledge is in proportion considerably wider. Every letter has its equivalent in other languages, and its equivalent in a figure or figures of the calculation table. It has also numerous other significations, which depend upon the special idiosyncrasies and characteristics of the person, object, or subject to be studied. As the Hindus claim to have received the Devanāgarī characters from Sarasvatī, the inventress of Sanskrit, the "language of the Devas" or Gods (in their exoteric pantheon), so most of the ancient nations claimed the same privilege for the origin of their letters and tongue. The *Kabalah* calls the Hebrew alphabet the "letters of the Angels," which were communicated to the Patriarchs, just as the Devanāgarī was to the Rishis by the Devas. The Chaldeans found their letters traced in the sky by the "yet unsettled stars and comets," says the *Book of Numbers*; while the Phoenicians had a sacred alphabet formed by the twisting of the sacred serpents. The Natar Khari (hieratic alphabet) and secret (sacerdotal) speech of the Egyptians is closely related to the oldest "Secret Doctrine Speech." It is a Devanāgarī with mystical combinations and additions into which the Senzar largely enters.[25]

Senzar—The Sacerdotal Secret Tongue

The Senzar and Sanskrit alphabets and other Occult tongues, besides other potencies, have a number, colour,

25. SD III:99-100; V:114-5, 6-vol. ed.

and distinct syllable for every letter, and so had also the old Mosaic Hebrew.[26]

Another reference is made to Senzar, but its significance is explained only by means of a quotation from the writings of the French Mason, Ragon, in his *Maçonnerie Occulte*:

> Truly there was a time when numbers and alphabetical characters meant something more than they do now—the images of a mere insignificant sound. "Their mission was nobler then. Each of them represented by its form a complete sense, which besides the meaning of the word, had a double interpretation adapted to a dual doctrine." (p. 431)
>
> The system of the so-called Senzar characters is still more wonderful and difficult, since each letter is made to yield several meanings, a sign placed at the commencement showing the true meaning.[27]

Since the forefathers of the Toltecs understood Senzar, it is evident that the knowledge of that occult language was widespread.

The word Toltecs itself signifies "master builders" and is applied to a race which inhabited the central part of Mexico prior to the Aztec culture. According to their traditions the Toltecs established themselves in the valley of Mexico about 500 A.D. and continued to exist there until 1100. Evidently the Toltecs migrated to Yucatan, while others moved southward into Guatemala, Honduras, Salvador, and Nicaragua. It is now believed that the Toltecs adopted the Nahuatl language. Their greatest monuments, preserved to this day, are at San Juan de Teotihuacan, Cholula, and Tezcoco. Teotihuacan is situated about thirty miles northeast of Mexico city. A ceremonial site was erected there measuring three and one-half by two miles, set aside for the construction of pyramids and temples. Cholula, situated on the plateau of Anahuac, 7000 feet above sea level and about fifty-five miles east-southeast of Mexico City, was formerly known as Chunultecol. In its heyday it had 400 temples and

26. Ibid. III.530; V:505, 6-vol. ed.

27. Ibid. III:103; V:117, 6-vol. ed.

a population of 100,000; at the time of Cortez (1519) the city had 20,000 houses. The pyramid of Cholula is an imposing structure; its base is twice the size of that of the pyramid of Cheops at Ghizeh.

Although the Aztecs replaced the Toltecs in the valley of Mexico and established their capital there in 1325, the Aztecs also adopted the Nahuatl language.

This account of the Toltecs is based on present day knowledge concerning the pre-Aztec culture. Nevertheless, in view of the fact that H. P. Blavatsky referred to the forefathers of the Toltecs as having knowledge of Senzar, one should ascribe a much greater antiquity to that earlier culture which flourished prior to the Toltecs. In fact, not only was Senzar referred to as being known in Atlantis and therefore was "the secret sacerdotal language or the 'Mystery-speech' of the initiated Adepts, all over the world,"[28] but as stated, Senzar was used by the sages of the Third Root Race, referred to as the Mānushis.

Mānushi is a Sanskrit adjective signifying "human." The word is also associated with Buddhas, especially in northern Buddhism. In this connotation the word signifies Racial Buddhas—regarded as Regents of Root-Races—in contradistinction to Dhyāni Buddhas, who are the Watchers or Regents of the Rounds. The term is also applied to "incarnated Dhyāni Chohans" of the Third Root-Race, as is the case in the quotation from *The Secret Doctrine* on the old Book. The "incarnated Dhyāni Chohans" are referred to in this śloka from the Stanzas of Dzyan:

> The Sons of Wisdom, the Sons of Night, ready for rebirth, came down. . . . "In these shall we dwell," said the Lords of the Flame.[29]

In this śloka the Sons of Wisdom and Sons of Night are synonymous terms, although more frequently referred to as the Mānasa-putras (Sons of Mind), or again Lords of the Flames—the "Flame" signifying Manas, the Mind-principle.

---

28. TG, 295.

29. SD Stanza VII, sloka 24, 2nd series.

These superior beings assisted in the evolutionary development of humanity during the Third Root Race. They are termed Sons of Night because of having attained their superior status in a prior manvantara.

The Lords of the Flame are the "incarnated Dhyāni Chohans." So are the *Devas* of the Second and First Root-Races, although the word here is used in a particular sense—that of superior beings assisting the humans of the earliest Races. Literally the word *deva* means a shining being; it is derived from the verb root *div,* to shine, and is usually rendered "a celestial being." In this particular case the Devas of the Second Root-Race signify the Asuras and Kumāras who incarnated in the Second Root-Race vehicles. To quote a Commentary from the Book of Dzyan:

> This is the Second Root-Race, endowed by the preservers (Rākshasas) and the incarnating gods (Asuras and the Kumāras) with the first primitive and weak spark (the germ of intelligence).[30]

But this "weak spark" was not sufficient to enlighten the humans of the Second Root-Race. It became the karmic requirement of the Mānasaputras—the Lords of the Flame—to fully enlighten the humans of the Third Root-Race, thus enabling the mind-principle to function consciously.

The Book of Dzyan and the Kiu-ti

Even though the passage quoted from *The Secret Doctrine* mentioned some of the works which had been derived from the archaic manuscript, it did not include the specific work which was preeminently used and which formed the basis for the writing of H. P. Blavatsky's work, namely, the *Book of Dzyan.* This book originated from the *Kiu-ti.* From this same source came the *Stanzas of Dzyan.* The word *Dzyan* is thus explained:

> *Dan,* now become in modern Chinese and Tibetan phonetics *Ch'an,* is the general term for the esoteric

30. Ibid. II:164-5; III:172, 6-vol. ed.; II:174, 3rd ed.

schools, and their literature. In the old books, the word *Janna* is defined as "to reform one's self by meditation and knowledge," a second inner birth. Hence Dzan, *Djan* phonetically, the "Book of *Dzyan*."[31]

The *Book of Dzyan*—from the Sanskrit word "Dhyan"[32] (mystic meditation)—is the first volume of the Commentaries upon the seven secret folios of Kiu-te,[33] and a Glossary of the public works of the same name. Thirty-five volumes of Kiu-te for exoteric purposes and the use of the laymen may be found in the possession of the Tibetan Gelugpa Lamas, in the library of any monastery; and also fourteen books of Commentaries and Annotations on the same by the initiated Teachers.

Strictly speaking, those thirty-five books ought to be termed "The Popularised Version" of the Secret Doctrine, full of myths, blinds, and errors; the fourteen volumes of *Commentaries,* on the other hand—with their translations, annotations, and an ample glossary of Occult terms, worked out from one small archaic folio, the *Book of the Secret Wisdom of the World*[34]—contain a digest of all the Occult Sciences. These, it appears, are kept secret and apart, in the charge of the Teshu Lama of Tji-gad-je.[35] The Books of Kiu-te are comparatively modern, having been edited within the last millennium, whereas, the earliest volumes of the *Commentaries* are

31. Ibid. I:xx; I:44, 6-vol. ed.; I:4, 3rd ed.

32. *Dhyāna,* derived from the verb root *dhyai,* to meditate. Cf. the symbolic words in *The Voice of the Silence*: "Dhyāna whose golden gate once opened leads the Naljor [a saint, an adept] toward the realm of Sat eternal and its ceaseless contemplation." (p. 48, orig. ed.)

33. Also spelled Kiu-ti. In the next sentence "Gelugpa Lamas" designate the "Yellow Cap Lamas," those Tibetan Lamas who adhere to the reforms in Buddhism instituted under the administration of Tsong-Kha-pa in 1417.

34. Here H. P. Blavatsky added the following footnote: "It is from the texts of all these works that *The Secret Doctrine* has been given. The original matter would not make a small pamphlet, but the explanations and notes from the Commentaries and Glossaries might be worked into ten volumes as large as *Isis Unveiled.*" SD III:405; V:389, 6-vol. ed.

35. Tji-gad-je (modernized spelling Shigatse) was the seat of the Tashi-Lama of Tibet; situated near the joining of the Nyang-Chu with the Tsang-po River. Southwest of Shigatse is the celebrated monastery Tashi-Lhünpo, founded by Ganden Truppa in 1445.

> of untold antiquity, some fragments of the original cylinders having been preserved. With the exception that they explain and correct some of the too fabulous, and to every appearance, grossly exaggerated accounts in the Books of Kiu-te—properly so-called—the *Commentaries* have little to do with these. They stand in relation to them as the Chaldaeo-Jewish *Kabalah* stands to the Mosaic Books.[36]
>
> No student, unless very advanced, would be benefited by the perusal of those exoteric volumes. They must be read with a key to their meaning, and that key can only be found in the Commentaries.[37]

H. P. Blavatsky then quotes Thango-pa Chhe-go-mo, a Tibetan Lama, in connection with the Tibetan Sacred Canon and the secret folios to which he had access:

> It is true that the Sacred Canon of the Tibetans, the Kaghyur [Kanjur] and Bstanhgyur [Tanjur], comprises 1707 distinct works—1083 public and 624 secret volumes, the former being composed of 350 and the latter of 77 volumes folio. May we humbly invite the good missionaries, however, to tell us when they ever succeeded in getting a glimpse of the last-named secret folios? Had they even by chance seen them I can assure the Western Pandits that these manuscripts and folios could never be understood even by a born Tibetan without a key (*a*) to their peculiar characters, and (*b*) to their hidden meaning. In our system every description of locality is figurative, every name and word purposely veiled; and one has first to study the mode of deciphering and then to learn the equivalent secret terms and symbols for nearly every word of the religious language. The Egyptian enchorial or hieratic system is child's play to our sacerdotal puzzles.[38]

There is a passage in *The Mahatma Letters to A. P. Sinnett* regarding the *Kiu-ti* which should be included at this

36. SD III:405-6; V:389, 6-vol. ed.
37. Ibid. III:406; V:390, 6-vol. ed.
38. Ibid.

point:

> The monad performs not only "world rings" or seven major inmetalliations, inherbations, zoonisations and incarnations—but an infinitude of sub-rings or subordinate whirls all in series of sevens. As the geologist divides the crust of the earth into great divisions, subdivisions, minor compartments and zones; and the botanist his plants into orders, classes and species, and the zoologist his subjects into classes, orders and families, so we have our arbitrary classifications and our nomenclature. But besides all this being incomprehensible to you, volumes upon volumes out of the Books of Kiu-te and others would have to be written. Their commentaries are worse still. They are filled with the most abstruse mathematical calculations the key to most of which are in the hands of our highest adepts only, since showing as they do the infinitude of the phenomenal manifestations in the side projections of the *one* Force they are again secret.[39]

This truly indicates that there is a vast treasure house of occult lore which is available only to one who is qualified to find it. It should be borne in mind that H. P. Blavatsky was selected to give certain portions of this occult lore in *The Secret Doctrine.* Yet regarding the immemorial Wisdom she wrote:

> . . . the now *Secret* Wisdom was once the one fountain head, the everflowing perennial source, at which were fed all its streamlets—the later religions of all nations—from the first down to the last.[40]

---

39. ML, (Letter No. 14) p. 81; p. 80, 3rd ed.

40. SD I: xliv-xlv; I:66, 6-vol. ed.; I:27-8, 3rd ed.

3.

ON THE ORIGIN OF THE SPARKS

IN THE opening chapter consideration was given to man's immortal component, referred to as the Pilgrim, or symbolically as a Spark of Eternity, thus calling attention to the divine aspect of man which issued forth as a spark from the eternal Flame. How or when did this spark originate? That, indeed, is the mystery of mysteries, which all accounts of genesis seek to portray in one way or another, most frequently by allegory or myth.

The Secret Doctrine portrays this even allegorically, principally by means of the Stanzas of Dzyan. The reference to the origin of life—specifically to the spark of eternity or monad—is presented in the first series of the Stanzas, which deal primarily with cosmogenesis. But it will be seen that the concept is applicable in either a macrocosmic or a microcosmic frame of reference. Actually the Stanzas of Dzyan, as set forth in *The Secret Doctrine* by H. P. Blavatsky, are extracts from the Book of Dzyan.[1] In turn the stanzas are subdivided into ślokas, a Sanskrit word often rendered "verses," although it has other meanings as well, such as a call or voice (as of the gods), praise, a hymn of praise, or even just a stanza. Because of the manner in which the Stanzas are written, it would be difficult to understand them were it not for the assistance provided by Mme. Blavatsky by means of parenthetical additions and subsequent explanatory passages. In addition, and perhaps of even greater importance, are the

---

1. An explanation regarding the Book of Dzyan and its origin was given in Chapter 2. It may be rendered "the Book of Meditation."

Commentaries on the Book of Dzyan about which H. P. B. wrote:

> In the apocalyptic ślokas of the Archaic Records, the language is as symbolical, if less mythical, than in the Purānas. Without the help of the later *commentaries,* compiled by generations of adepts, it would be impossible to understand the meaning correctly.[2]

The first śloka of Stanza VII (first series) is presented as though a guru were instructing his disciple (lanoo or chela):

> Behold the beginning of sentient formless life. First, the divine (vehicle), the one from the Mother-spirit (Ātman); then the Spiritual (Ātma-Buddhi, Spirit-soul); (again) the three from the one, the four from the one, and the five, from which the three, the five and the seven—these are the three-fold and the four-fold downward; the "Mind-born" sons of the first Lord (Avalokiteśvara) the Shining Seven (the "Builders"). It is they who are thou, me, him, O Lanoo; they who watch over thee and thy mother, Bhūmi (the Earth).[3]

Can the human mind conceive of "formless life"? of life without a form? The whole concept of life in the western world is so attached to and rooted in the form-side of things that one becomes bewildered when told to regard life without a form. H. P. Blavatsky explains that the first emanations following that from the "Mother-spirit," termed the "three from the one,"

> . . . were without form or substance (our *arūpa* triad), as neither the celestial nor the terrestrial substance had yet differentiated, "nor had the essence of things been formed."[4]

Of course, the first emanation of the three-in-one (often represented by the emanation of the three Logoi) is a concept

---

2. SD II:23; III;36, 6-vol. ed.; II:26, 3rd ed.

3. Ibid. I:213; I:260, 6-vol. ed.; I:233, 3rd ed.

4. Ibid. I:214; I:261, 6-vol. ed.; I:234, 3rd ed.

usually associated with cosmogenesis, nevertheless the initial stages pertaining to the microcosm would be along parallel lines: the three from the one representing Ātma-Buddhi-Manas, the three highest principles in the human constitution, Manas denoting the mind principle. Continuing the enumeration: the four includes Kāma, the energizing principle; the five, Prāna, the vital principle; the six and seven, Linga-śarīra, the model vehicle, and Sthūla-śarīra, the physical vehicle mentioned in the second śloka:

> The one ray multiplies the smaller rays. Life precedes form, and life survives the last atom (or Form, Sthūla-śarīra, external body). Through the countless rays proceeds the life-ray, the one, like a thread through many beads (pearls).[5]

The Sūtrātman and the Spark

The one life-ray (Ātman), proceeding like a thread, is termed the Sūtṛātman, the thread-soul (sūtra, a thread), which in the cosmical sense knits the universe together. This again stresses the concept of the One Life, as was done in the last sentence of the first śloka. However, when connected with humanity, the Sūtrātman signifies the continuity of existence from one cycle to another. For even though death may terminate a cycle, or period of life on earth, the Sūtrātman bridges the gulf of death—the latter having been caused when the external form is vacated by the immortal triad. But the thread-soul is not severed in the process, since life had already preceded the external form (the Sthūla śarīra). In due time the Sūtrātman is linked to another external vehicle for another period of life on earth. But this too will be ended; so will another, and yet another. The resultant life-period signifies the birth of a new person, because a new personality comes into manifestation. Yet each personality which has been left is not lost: it remains as a bead or pearl on the thread provided by Ātman. Furthermore, the attainments achieved by each personality (representing a lifetime

---

5. Ibid. I:222; I:269, 6-vol. ed; I:242, 3rd ed.

on earth) are not lost, because the life-periods are linked together by means of the Sūtrātman. Thus the present personality, although a new one, is the resultant not only of the previous person, but of all the former personalities.

The third śloka of Stanza VII definitely relates to the human kingdom.

> When the one becomes two—the "three-fold" appears. The three are (linked into) one; and it is our thread, O Lanoo, the heart of the man-plant, called Saptaparna.[6]

As with many of the ślokas, the first sentence may be given a cosmic as well as a human interpretation, signifying the manifestation of the three Logoi, as already considered in connection with the first śloka. An additional exposition is now added:

> When the One becomes two, the three-fold appears: to wit, when the One Eternal drops its reflection into the region of Manifestation, that reflection, "the Ray," differentiates the "Water of Space" [or Chaos]. . . . Chaos becomes male-female, and Water, incubated through Light, and the "three-fold being issues as its First-born."[7]

The second sentence gives one of the symbolical terms used in the Esoteric Philosophy for man: Saptaparna, here rendered "man-plant," for the reason that *parna* means "a leaf," and *sapta,* "seven," signifying thereby the seven principles of the human constitution. The three principles linked into one are Ātma-Buddhi-Manas; "our thread" signifies the Sūtrātman. The significance of "our thread" is further elaborated in the sixth śloka:

> From the first-born (primitive, or the first man) the thread between the silent watcher and his shadow becomes more strong and radiant with every change (reincarnation).[8]

---

6. Ibid. I:231; I:277, 6-vol. ed.; I:251-2, 3rd ed.

7. Ibid.

8. Ibid. I:264-5; I:308, 6-vol. ed.; I:285, 3rd ed.

The silent watcher stands for the Monad; his shadow, the evanescent personality. H. P. Blavatsky added this explanation:

> The Watcher, or the divine prototype, is at the upper rung of the ladder of being; the shadow at the lower. . . . Its *Primary,* the Spirit (Ātman) is one, of course, with Paramātman (the one Universal Spirit), but the vehicle (Vāhana) it is enshrined in, the Buddhi.[9]

Mention was made in one of the verses to the thread of life as "the thread of Fohat": "The spark hangs from the flame by the finest thread of Fohat." (Stanza VII, śloka 5) The term was used in a specific sense and is best explained by quoting a passage from *The Secret Doctrine,* which also describes the process of the coming into being of the Spark, or Monad, by means of uniting the sixth principle (Buddhi) to the seventh (Ātman), thus linking the Spark and the Flame. The paragraph is added as an interpretation of another śloka:

> Fohat traces spiral lines to unite the sixth to the seventh—the Crown. (Stanza V, śloka 4, first series)
>
> This tracing of "spiral lines" refers to the evolution of man's as well as Nature's principles; an evolution which takes place gradually, as does everything else in nature. The Sixth principle in man (Buddhi, the Divine Soul) though a mere breath, in our conceptions, is still something material when compared with divine "Spirit" (Ātman) of which it is the carrier or vehicle [upādhi]. Fohat, in his capacity of DIVINE LOVE (*Eros*), the electric Power of affinity and sympathy, is shown allegorically as trying to bring the pure Spirit, the Ray inseparable from the ONE absolute, into union with the Soul, the two constituting in Man the MONAD, and in Nature the first link between the ever unconditioned and the manifested.[10]

---

9. Ibid. I:265; I:308, 6-vol. ed.; I:285, 3rd ed.

10. Ibid. I:119; I:178, 6-vol. ed.; I:144, 3rd ed.

Referring again to the Thread of Fohat, H. P. Blavatsky asked the question "What is that 'Spark' which 'hangs from the Flame'?" and answered:

> It is Jīva, the Monad in conjunction with Manas, or rather its aroma—that which remains from each personality, when worthy, and hangs from Ātma-Buddhi, the Flame, by the thread of Life.[11]

It is Manas, the Mind-principle, which enables the Monad to project its "thread of life" to the lower vehicles, for it is Higher Manas (technically Manas-Taijasī, that is Manas with the light of Buddhi, here likened to the "aroma of Manas") which is also termed the Reincarnating Ego.

The Day Be With Us and Paranirvāna

The subject presented in the seventh śloka unfolds an entirely different viewpoint from the preceding narrative, even though still concerned with the Sparks. Instead of the *origin* of the Sparks from the One or the Source, consideration is given to the return or the mergence of the Spark with its Source. Nevertheless, it is in keeping with the vast scope of the Divine Plan and its fundamental law—the law of periodicity, enunciated in the second fundamental proposition of *The Secret Doctrine*: for every period of activity there is a consequent period of rest. Since the period of the origin of the Sparks is under consideration, there is bound to be a consequent period for the return of the Sparks. The śloka is presented in the form of a dialogue between the Flame and the Spark.

> "This is thy present Wheel"—said the Flame to the Spark. Thou art myself, my image and my shadow. I have clothed myself in thee, and thou art my vāhana (vehicle) to the day, 'Be with us,' when thou shalt rebecome myself and others, thyself and me." (Stanza VII, śloka 7)[12]

---

11. Ibid. I:238; I:284, 6-vol. ed.; I:259, 3rd ed.

12. Ibid. I:265; I:309, 6-vol. ed.; I:286, 3rd ed.

The expression "present wheel" refers to a globe or sphere, such as our Earth. In this śloka the Flame signifies the Monadic Essence, Ātman, the Universal Spirit, from which the Spark arises; the Spark is the Pilgrim, the Monad, Ātma-Buddhi, the manifested spiritual essence. The first three words of the second sentence: "Thou art myself" (which are the words of the Flame) recall to mind the opening statement of the third fundamental proposition of *The Secret Doctrine*: "The fundamental identity of all Souls with the Universal Over-Soul, the latter being itself an aspect of the Unknown Root."[13] Further, to impress the importance of the idea, the words "my image" are added, signifying the Spark. The words "my shadow" represent the personality—an additional aspect provided to the image, implying that the Spark is thereby darkened because of its connection with a vehicle (vāhana). The very next sentence in fact explains this idea: "I have clothed myself in thee, and thou art my vāhana (vehicle) to the day 'Be with us.'" How very expressive are these three final words! No longer is the concept conveyed in the singular; instead the all-embracing figure of speech, "us" is used. One aspect of the Day Be With Us signifies the Mahā-pralaya—the Very Great Night.

Can the full significance of the term "the Day Be With Us" be grasped by the finite mind? The prime function of the mind is to objectify and this very fact limits its outlook; the mind identifies itself with a single, limited point of view. The salient feature of the Day Be With Us is the concept of freedom from bonds as well as bounds of limitation. So the term signifies that the Many will have rebecome the One, for the reason that no longer will there be a need for vehicles (vāhanas), because a Mahā-pralaya encompasses the manifested cosmic system—that is to say the sun, as well as all the planets of the system and all that is in them.

The meaning of Mahā-pralaya should be explained: *mahā* means great; *pralaya,* a Sanskrit compound—*pra,* away; *laya* from the verb-root *lī,* to dissolve; hence "dissolution" of forms, but not of the primordial essence of substance itself

---

13. Ibid. I:17; I:82, 6-vol. ed.; I:45, 3rd ed.

(termed Prakriti), which remains in a state of latency, of homogeneity. A Mahā-pralaya is also termed a Prākritika-pralaya, when everything that exists is resolved into the primal element (Prakriti). Prākritika is an adjective denoting pertaining to Prakriti, primordial substance. This state of dissolution lasts for a Mahā-kalpa, which signifies a Great Age—a period of time equivalent to the corresponding Age or Life of Brahmā, which is said to endure for a period of 311,040,000,000,000 solar years and in our terms is the life of the solar system.

The words of the śloka have expressed the idea conveyed in the term Day Be With Us symbolically: "Thou shalt re-become myself and others, thyself and me," for the spark shall no longer be as a luminous particle dimmed by its vehicle, but shall be as the very Flame itself, because of its restored unity with the Flame.

However, it should be borne in mind that this is not "the end." For after a Mahā-kalpa the spark will once again re-emerge, even though during the Mahā-pralaya "the dewdrop has slipped into the shining sea," as another poetic version describes the process of existence in an ever-cycling alternating state of being and non-being. This is the esoteric explanation of Nirvāna, although the term for the state of the monads during the state of the Day Be With Us is known as Paranirvāna, the highest Nirvāna. Yet Paranirvāna, or Paranishpanna, is a state of absolute perfection, a perfectly subjective condition of pure spirituality, which is as far superior to Nirvāna as that state is superior to Devachan.[14]

It has been observed by Orientalists that Nirvāna literally means "blown out." But what is it that is "blown out"? As explained in the ancient scripture of the Orient, the *Bhagavad-Gītā,*[15] Ātman is not extinguished when the body dies, for it is not affected by death; it is the body (the vāhana) that is "blown out," not the spark, for the spirit is no longer with the body. On the other hand, during the state of Nirvāna,

---

14. Cf. ML p.199; pp. 196-7, 3rd ed.

15. Chapter ii, ślokas 19-21, 29-30.

Ātman rebecomes united with Brahman, the Supreme always unmanifest Spirit.

The Secret Doctrine has this comment on the state of the Day Be With Us and Nirvāna:

> The day when "the spark will re-become the Flame (man will merge into his Dhyān Chohan) myself and others, thyself and me," as the Stanza has it—means this: In *Paranirvāna*—when *Pralaya* will have reduced not only material and psychical bodies, but even the spiritual *Ego(s)* to their original principle—the Past, Present, and even Future Humanities, like all things, will be one and the same. Everything will have reentered the *Great Breath.* In other words, everything will be "merged in Brahma" or the divine unity.[16]
>
> To see in Nirvāna annihilation amounts to saying of a man plunged in a sound *dreamless* sleep—*one that leaves no impression on the physical memory and brain, because the sleeper's Higher Self is in its original state of absolute consciousness* during those hours—that he, too, is annihilated. The latter simile answers only to one side of the question—the most material; since *re-absorption* is by no means such a "dreamless sleep," but, on the contrary, *absolute* existence, an unconditioned unity, or a state, to describe which human language is absolutely and hopelessly inadequate . . . . Nor is the individuality—*nor even the essence of the personality*, if any be left behind—lost, because re-absorbed. For, however limitless—from a human standpoint—the paranirvānic state, it has yet a limit in Eternity. Once reached, the same monad will *re-emerge* therefrom, as a still higher being, on a far higher plane, to recommence its cycle of perfected activity.[17]

In further explanation of the Day-Be-With-Us, H. P. Blavatsky relates how the ancient Egyptians expressed the same idea, for they

> . . . called the same the "Day of Come-to-Us," which is identical with the former, though the verb "be," in this

16. SD I:265-6; I:309, 6-vol. ed.; I:286, 3rd ed.

17. Ibid. I:266; I:309, 6-vol. ed.; I:286-7, 3rd ed.

> sense, might be still better replaced with either of the two words "Remain" or "Rest-with-us," as it refers to that long period of REST which is called Paranirvāna. As in the exoteric interpretation of the Egyptian rites the soul of every defunct person—from the Hierophant down to the sacred bull Apis—became an Osiris, was Osirified, though the Secret Doctrine had always taught that the real Osirification was the lot of every Monad only after 3,000 cycles of Existences; so in the present case. The "Monad," born of the nature and the very Essence of the "Seven" (its highest principle becoming immediately enshrined in the Seventh Cosmic Element), has to perform its septenary gyration throughout the Cycle of Being and forms, from the highest to the lowest; and then again from man to God. At the threshold of Paranirvāna it reassumes its primeval Essence and becomes the Absolute once more.[18]

It should be borne in mind that in the last sentence the words "it reassumes" signify that the Monadic Essence merges with the Absolute or Parabrahman—the latter meaning the boundless reaches of Infinitude. A distinction should be made between the Monad (Ātma-Buddhi) and the Monadic Essence (Ātman). In thus reassuming its primeval Essence the Monadic Essence no longer functions by means of its upādhis or vestures, because it is no longer functioning in the planes of manifestation. The planes of manifestation signify the four lower cosmic planes: they are numbered in descending sequence, the fourth, fifth, sixth, and seventh cosmic planes, in which the seven globes of the Earth planetary system have their being.

The Great Breath is a term which expresses that which is ceaseless, an endless process which may be regarded as the universal and eternal perpetual rhythmic motion. An inbreathing represents a period of rest—of non-manifestation, a pralaya. An outbreathing represents a period of activity—of creativity, a manvantara. The Great Breath is regarded as one of the aspects of the Absolute, namely Motion; the two other aspects, which form a trinity, are Boundless Space

18. Ibid. I:134-5; I:192, 6-vol. ed.; I:159-60, 3rd ed.

and Duration. In *The Secret Doctrine* "duration" signifies "continuance of time," meaning that state of time which is neither affected by the alternating periods of activity and rest. The concept of time is based upon the sequence of cycles, whether great or small, which have a beginning and an ending. In connection with the beginning of a system—its coming into being—whether planetary or solar, this is signified by the term "the Divine Breath."

> When the "Great Breath" is projected, it is called the Divine Breath, and is regarded as the breathing of the Unknowable Deity—the One Existence—which breathes out a thought, as it were, which becomes the Kosmos. So also is it when the Divine Breath is inspired again the Universe disappears into the bosom of "the Great Mother," who then sleeps "wrapped in her invisible robes."[19]

The Luminous Sons of Manvantaric Dawn

The concluding portion of the seventh śloka has a direct reference to the Peopling of the Earth. As may be expected it is presented in allegorical manner. The idea of "creation" is introduced through the "Builders." "Then the Builders having donned their first clothing . . ." (Stanza VII, śloka 7) "Donning their first clothing" signifies adapting themselves to forms suitable to each one of the seven globes of the planetary system—even though they are regarded as "arūpa" beings, that is to say, beings with "not a form" of physicalized substance. The Builders are termed the Luminous Sons of Manvantaric Dawn and are referred to as the real creators of the Universe. However, it should be pointed out that there are two classes which may be designated as "Builders," namely the Architects and the Constructors. The Architects are regarded as the designers, whereas the Constructors are the producers of forms. In connection with the formation of a universe, or a cosmos, two classes are involved: (1) the beings who represent the intelligence aspect, usually referred

19. Ibid. I:43; I:115-16, 6-vol. ed.; I:74, 3rd ed.

to in *The Secret Doctrine* as the Watchers, or Regents of a system. They are also termed Architects and Dhyānis, inasmuch as they function in a supervisory capacity. (2) The beings associated with producing forms, customarily regarded as creators, or the builders of forms, termed generally Dhyāni-Chohans. Thus they furnish forms for all the Kingdoms of Nature.

In the quotation which follows, the Builders are referred to as the Architects of the Planetary System and are also called the Watchers of the Seven Spheres. Observe that the Seven Spheres are also regarded in a dual capacity: (1) as the Seven Sacred Planets; (2) as the seven Globes of a planetary system, which are usually referred to as a planetary chain (or a chain of seven globes) —here they are mentioned as seven spheres or planets.

> The "Builders," the "Sons of Manvantaric Dawn," are the real creators of the Universe; and in this doctrine, which deals only with our Planetary System, they, as the architects of the latter, are also called the "Watchers" of the Seven Spheres, which exoterically are the Seven planets, and esoterically the seven earths or spheres (planets) of our chain also.[20]

The śloka continues: The Builders "descend on radiant earth, and reign over men." And the reigning is thus described:

> The "Watchers" reign over man during the whole period of *Satya Yuga* and the smaller subsequent yugas, down to the beginning of the Third Root Race; after which it is the Patriarchs, Heroes, and the Manes . . . the incarnated Dhyānis of a lower order.[21]
>
> The Secret Doctrine teaches . . . that the Dhyāni-Buddhas of the two higher groups, namely, the "Watchers" or the "Architects," furnished the many and various races with divine kings and leaders. It is the latter who taught humanity their arts and sciences,

---

20. Ibid. I:53; I:124, 6-vol. ed.; I:83, 3rd ed.

21. Ibid. I:266; I:310, 6-vol. ed.; I:287, 3rd ed.

and the former who revealed to the incarnated Monads that had just shaken off their vehicles of the lower Kingdoms—and who had, therefore, lost every recollection of their divine origin—the great spiritual truths of the transcendental worlds.

Thus, as expressed in the Stanza, the Watchers descended on Earth and reigned over men—"*who are themselves.*" The reigning kings had finished their cycle on Earth and other worlds, in the preceding Rounds. In the future manvantaras they will have risen to higher systems than our planetary world.[22]

The Four Yugas

Even from the earliest times of our planet's history, the Earth has been under the supervision of the Watchers. Therefore, the Yugas and Root-Races are included in this regency. Thus the Satya-yuga mentioned in the quotation had reference to the initial period of the peopling of the Earth—termed the First Root-Race—a time-period of several million years prior to the Third Root-Race which was specified. The Satya-yuga signifies "the Age of Truth"; it is also called the Krita-yuga, "the Age that is well accomplished." In addition it is described as the Golden Age and indicates the initial cyclic period of a Root-Race. The time-period of the Satya-yuga is equivalent to 1,728,000 years. The "smaller subsequent yugas" refer to the three yugas which follow the initial cyclic period. The subsequent yugas are named:

The Tretā-yuga
– "the age of the 3 Fires" lasting 1,296,000 years
The Dvāpara-yuga
– "the Second Age" lasting 864,000 years
The Kali-yuga
– "the Dark Age" (the Iron Age) lasting 432,000 years

There is yet one more aspect of the śloka to be considered, for it was stated that "the Builders descend on radiant earth."

---

22. Ibid. I:267; I:310, 6-vol. ed.; I:287, 3rd ed.

The Earth as we know it now was not always so, especially when reference is made to a period of time "covering hundreds of millions of years ago." In fact one of the Commentaries on the Book of Dzyan explains:

> The globe was "*fiery, cool and radiant* as its ethereal men and animals during the first Round," says the Commentary, uttering a contradiction or paradox in the opinion of our present Science; "*luminous* and more dense and heavy–during the second Round; *watery* during the Third."[23]

This is further explained by H. P. Blavatsky:

> This means that every new Round develops one of the Compound Elements, as now known to Science . . . they have to evolve, progress, and increase to the Manvantaric end. Thus the First Round, we are taught, developed but one Element, and a nature and humanity in what may be called one aspect of Nature–called by some, very unscientifically, though it may be so *de facto,* "One-dimensional Space." . . .
>
> The succession of primary aspects of Nature with which the succession of Rounds is concerned, has to do, as already indicated, with the development of the "Elements" (in the Occult sense)–Fire, Air, Water, Earth. We are only in the fourth Round, and our catalog so far stops short. The centers of consciousness (destined to develop into humanity as we know it) of the third Round arrived at a perception of the third Element, Water. Those of the fourth Round have added *earth* as a state of matter to their stock as well as the three other elements in their present transformation.[24]

In Sanskrit these Elements are termed Tattvas, and their Sanskrit names, in corresponding sequence are: Taijasa-tattva, Fire; Vāyu-tattva, Air; Apas-tattva, Water; Prithivī-tattva, Earth.

One final quotation informs us concerning the origin of the Sparks (here called Beams or Monads), and it also shows

---

23. Ibid. I:252 fn.; I:297, 6-vol. ed.; I:273, 3rd ed.

24. Ibid. I:250-53; I:295-7, 6-vol. ed.; I:270-3, 3rd ed.

the link between the Sparks and the Flame (here called the Breath):

> . . . it is . . . the "Seven Breaths," so called, that furnish man with his immortal Monad in his cyclic pilgrimage.
>
> Descending on his region first as Lord of Glory, the Flame (or Breath), having called into conscious being the highest of the Emanations of that special region, ascends from it again to Its primeval seat, whence it watches over and guides its countless Beams (Monads) . . . It overshadows each with one of Its countless beams . . . Yet even the "beam" is a part of the Lord of Lords—the title of the highest Dhyān-Chohans.[25]

25. Ibid. III:369-70; V:357, 6-vol. ed.

4.

THE PLANETARY REGENTS

TWO SERIES of the Stanzas of Dzyan are presented in *The Secret Doctrine.* One considers the coming into being of a system, whether it be a universe, a solar system, or a planetary system. The second gives an account of the peopling of the earth. It is presented symbolically, and its opening words describe the beings who are the regents of the planetary system. So an introductory explanation of the symbols will no doubt be helpful.

In the Ageless Wisdom all the heavenly bodies are regarded as "living beings." The very word Zodiac indicates this. It is derived from the Greek word *zoion,* signifying an animal. Furthermore, the original meaning of the Latin word *anima* signifies a living (animated) being. This idea runs through the ancient Stanzas. Regarding the zodiac itself, H. P. Blavatsky wrote:

> The Zodiac was known in India and Egypt for incalculable ages, and the knowledge of the sages (magi) of these countries, with regard to the occult influence of the stars and heavenly bodies on our earth, was far greater than profane astronomy can ever hope to reach to. If, even now, when most of the secrets of the Asuramayas and the Zoroasters are lost, it is still amply shown that horoscopes and judiciary astrology are far from being based on fiction, and if such men as Kepler and even Sir Isaac Newton believed that stars and constellations influenced the destiny of our globe and its humanities, it requires no great stretch of faith to believe that men who were initiated into all the mysteries

of nature, as well as into astronomy and astrology, knew precisely in what way nations and mankind, whole races as well as individuals, would be affected by the so-called "signs of the Zodiac."[1]

*Why the Earth Is Regarded As the Fourth Sphere**

Another basic concept is stated as a postulate: "Everything in the metaphysical as in the physical universe is septenary."[2] This is not only applicable to beings existing upon the earth; it applies to the planet itself. Thus in the familiar description of the composition of man, humanity is defined as composed of seven principles, six of which are subjective, while only the seventh is objective, namely the physical body. In like manner the earth is the objective sphere of the earth planetary system; for it is accompanied by six subjective spheres or globes. The six companion globes are termed subjective because they are situated on cosmic planes which are superior to the physical cosmic plane on which the Earth is stationed. The Earth is designated by the fourth letter of the alphabet, D, whereas the six other globes are enumerated in this manner:

Globes A and G are situated on the 4th cosmic plane;
Globes B and F are situated on the 5th cosmic plane:
Globes C and E are situated on the 6th cosmic plane;
Globe D is situated on the 7th cosmic plane.

The principal reason that our Earth is designated the fourth of the sevenfold globes is because, on the descending arc, physical evolution reaches its lowermost point on this fourth globe. The Descending Arc signifies evolution on Globes A, B, C, D. The Ascending Arc represents evolution on Globes E, F, G. Each of the four manifested kingdoms—namely the Human Kingdom, the Animal Kingdom, the Plant Kingdom, the Mineral Kingdom—undergoes its own

---

* This material was presented briefly in Chapter 1. It is here discussed in more detail.

1. TG 387-8.

2. SD I:158; I:213, 6-vol. ed.; I:182, 3rd ed.

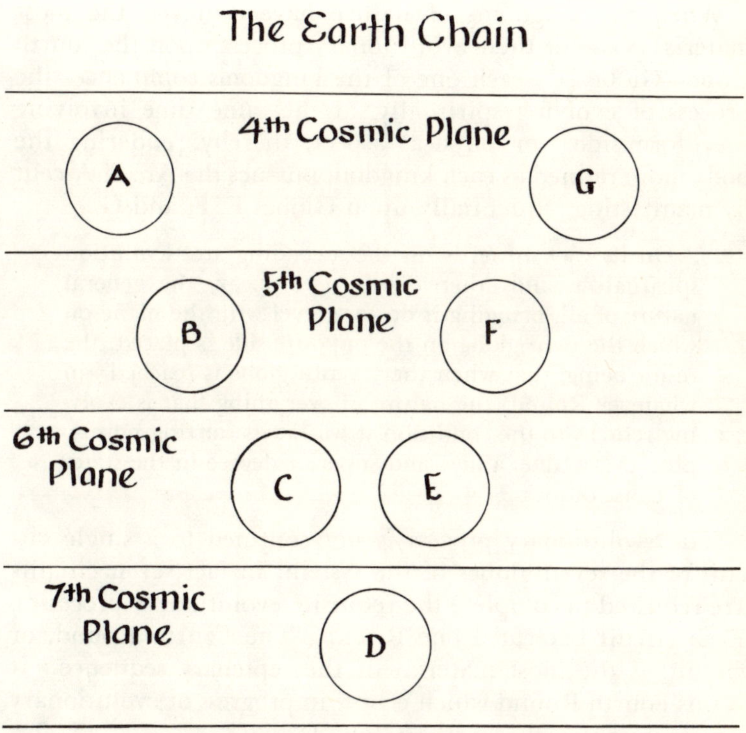

DIAGRAM 4

particular scheme of evolutionary development. Nevertheless, the four manifested kingdoms of nature are described as being involved in a twofold evolutionary process: this consists of an evolution of forms (rūpas) or bodies, in which the body becomes progressively more material while a consequent involution of spirit takes place. This dual process of evolution is termed an Arc of Descent (or the Descending Arc) and an Arc of Ascent (or the Ascending Arc). The Arc of Descent consists in following a sequence of periods of activity in which forms are established upon each of the four first spheres or globes of the planetary system, thus evolving progressively more material or physicalized bodies (rūpas) on the globes designated as A, B, C, D. Globe D represents the most physical sphere of the series.

When the kingdoms of nature have achieved the most material phase of their evolutionary process upon the fourth globe (Globe D), each one of the kingdoms commences the process of evolving spiritually, at the same time involving the "form-side," or physical aspect, thereby rendering the body more refined as each kingdom pursues the Arc of Ascent by manifesting sequentially upon Globes E, F, and G.

> On its way upwards on the ascending arc, Evolūtion spiritualizes and etherealizes, so to speak, the general nature of all, bringing it on to a level with the plane on which the twin globe on the opposite side is placed; the result being, that when the seventh globe is reached (in whatever Round) the nature of everything that is evolving returns to the condition it was at its starting point—plus, every time, a new and superior degree in the states of consciousness.[3]

The evolutionary process is *not* confined to a single circuit of the seven globes of the system; in fact seven circuits are required to complete the requisite evolutionary processes. Each circuit is termed one Round. The Fourth Round, or circuit, is the most material of the septenary sequence. It is this Fourth Round which is now in progress of evolutionary development on the Earth (Globe D).

> In the middle of the Fourth revolution, which is our present "Round," Evolution has reached its acme of physical development, crowned its work with the perfect physical man, and, from this point, begins its work spiritward.[4]

Various Grades of Regents

In the very first śloka of the Stanzas of Dzyan, mention is made of the seven globes of a planetary system, with particular reference to the fourth sphere, our Earth. The second

---

3. Ibid. I:232; I:278, 6-vol. ed.; I:253, 3rd ed.

4. Ibid.

word of the verse names the great being who presides over the Earth:

> The Lha which turns the fourth is subservient to the Lha of the Seven, they who revolve driving their chariots around their Lord, the One Eye. His breath gave life to the Seven; it gave life to the first. (Stanza I, śloka 1, 2nd series)[5]

Lha, is a Tibetan word meaning a Planetary Spirit, and the "body"—the planet, the globe, the sphere of activity for all its living beings—is termed a "chariot," driven by the Regent, also named Rector or Watcher, who is indeed the Lha of the planet. In other ślokas the globe is referred to as a palace, signifying the domain of the planetary ruler. It should be mentioned that the word *lha* is a generalizing term, frequently used in Tibet and in exoteric usage is rendered a "god." H. P. Blavatsky explains:

> Lha is the ancient word in trans-Himalayan regions for "Spirit," any celestial or *superhuman* Being, and it covers the whole series of heavenly hierarchies, from Archangel, or Dhyāni, down to an angel of darkness, or terrestrial Spirit.[6]

The grade or status of a Lha is usually indicated by means of an adjective. In the opening sentence, the śloka indicates that the first Planetary Spirit referred to is the Regent of the fourth Globe, Globe D of the Earth-Chain; likewise there is a greater Rector who has regency over all the seven globes of the series.

> Again, it must be explained and remembered that as the work of each Round is said to be apportioned to a different group of so-called "Creators" or "Architects," so is that of every globe; i.e., it is under the supervision and guidance of special "Builders" and "Watchers"—the various Dhyan-Chohans.[7]

---

5. Ibid. II:15; III;27, 6-vol. ed.; II:15, 3rd ed.

6. Ibid. II:22; III:35, 6-vol. ed.; II:25, 3rd ed.

7. Ibid. I:233; I:278-9, 6-vol. ed.; I.253, 3rd ed.

A distinction should be made between the "Architects" and the "Builders"; the former represent the "intelligence or design aspect" of a planetary system in contradistinction to the "form or building aspect" of a Chain of seven globes, inasmuch as the function of the "Builders" is to provide forms (rūpas) for the various kingdoms, pursuant to the plans (or "blueprints") provided by the "Architects." Each kingdom has its own Watcher or Regent.

The compound word *Dhyan-Chohan* is a generalizing term signifying any being superior in status or grade to the Human Kingdom, hence regarded as *superhuman. The Secret Doctrine* places the Dhyan-Chohans in three Classes or Kingdoms, but does not designate their specific functions. The first word of the compound is an abbreviated form of the Sanskrit *dhyāni,* derived from the verb-root *dhyai,* to meditate; the second component, *chohan,* is a Tibetan word signifying lord; hence the term may be rendered "Meditating Lords."

There are varying grades of Regents, or Watchers. The most superior Regents of a planetary system are termed *Dhyāni-Buddhas,* who have regency over the Rounds of a chain of seven globes—one Dhyāni-Buddha for each of the Rounds. Dhyāni-Buddha is a northern Buddhistic term; in the Brahmanical scriptures the Manus are described as the Regents of a planetary system. The Watchers over the Globes are designated as Dhyāni-Bodhisattvas; the Regents over the Root-Races are known as Super-Terrestrial Bodhisattvas, who are assisted by Mānushya-Buddhas—a term signifying Human Buddhas.

In the opening words of the Stanza: "The Lha which turns the fourth," signifies the Dhyāni-Bodhisattva who has regency over our Earth (Globe D), the fourth globe of the planetary system.

The part of the first sentence of the Stanza: "the fourth is subservient to the Lha of the seven," is explained in *The Secret Doctrine*:

> This expression shows in plain language that the Spirit-Guardian of our globe, which is the fourth in the

> chain, is subordinate to the chief Spirit (or God) of the Seven Planetary Genii or Spirits.[8]

These are also referred to as the seven Dhyānis; specifically they are the Regents of the Seven Sacred Planets.[9]

The next phrase of the first sentence of the śloka reads: "they who revolve, driving their chariots around their Lord." This statement shows that the writers of this ancient manuscript had knowledge of the fact that the planets rotate on their axes as well as circle around the sun. Furthermore, the sun was regarded as their "Lord," and was entitled "one eye" (Eka-Chakshu in Sanskrit). Another of the many names given to the sun was the compound Sanskrit term Loka-Chakshu—"eye of the world." The phrase "the Seven Eyes of the Lord" is also met with in Christian terminology and the following explanation is provided:

> The ancients had, in their Kyriel of gods, seven chief Mystery-gods, whose chief was, *exoterically,* the visible Sun, or the eighth, and, *esoterically,* the *second Logos,* the Demiurge. The seven (who have now become the "Seven Eyes of the Lord" in the Christian religion) where the regents of the seven *chief* planets; but these were not reckoned according to the enumeration devised later by people who had forgotten, or who had an inadequate notion of, the real *Mysteries,* and included neither the sun, the moon, nor the earth. The sun was the chief, exoterically, of the twelve great gods, or zodiacal constellations; and, esoterically, the Messiah, the Christos (the subject *anointed* by the GREAT BREATH, or the ONE) surrounded by his twelve subordinate powers, also subordinate, in turn, to each of the seven "Mystery-gods" of the planets.[10]

The Sun Is the Giver of Life

The second sentence of the first śloka is a short, succinct statement: "His breath gives life to the seven." This has a

8. Ibid. II:22; III:35, 6-vol. ed.; II:25, 3rd ed.

9. An exposition of this significant passage will be given in the next chapter.

10. SD II:22-3; III:35-6, 6-vol. ed.; II:26, 3rd ed.

dual significance: (1) the sun's breath—a literal rendering of the Sanskrit word prāna, meaning "the breath of life" —gives life literally to the seven globes of the earth-chain; (2) the sun's prāna gives life to the Seven Sacred Planets (see Chapter 5).

The sun is often referred to in ancient scriptures as "the giver of life." This was expressed in a remarkable passage in a Commentary from the Book of Dzyan:

> Sūrya (the Sun) in its visible reflection, exhibits the first, or lowest state of the seventh, the highest state of the Universal PRESENCE, the pure of the pure, the first manifested Breath of the ever Unmanifested SAT (Be-ness). All the Central physical or objective Suns are in their substance the lowest state of the first Principle of the BREATH. . . .
>
> The real substance of the concealed (Sun) is a nucleus of Mother substance. It is the heart and the matrix of all the living and existing Forces in our solar universe. It is the Kernel from which proceed to spread on their cyclic journeys all the Powers that set in action the atoms in their functional duties, and the focus within which they again meet in their seventh essence every eleventh year.[11]

The "atoms" referred to are in esoteric terminology "life-atoms," the essential element of life or energy in an atom.

> Every atom and molecule in the Universe is both *life-giving* and *death-giving* to that form, inasmuch as it builds by aggregation universes and the ephemeral vehicles.[12]

The reference to every eleventh year brings to mind the eleven-year sun-spot cycle, "which is due to the contraction of the Solar Heart." The seventh essence is the Primordial Mother-Substance, also referred to as the inner essence, which acts as the "inner" kernel around which the "outer" substance or matter comprising a globe is concreted, whether a planet or a sun.

---

11. Ibid. I:289-90; I:330-1, 6-vol. ed.; I:309-10, 3rd ed.

12. Ibid. I:261; I:305, 6-vol. ed.; I:281, 3rd ed.

Resuming consideration of the sentence "His breath gives life to the seven," in addition to the two meanings already considered, H. P. Blavatsky supplies a third meaning of even greater importance:

> [it] refers as much to the sun, who gives life to the Planets, as to the "High One," the Spiritual Sun, who gives life to the whole Kosmos. The astronomical and astrological keys opening the gate leading to the mysteries of Theogony can be found only in the later glossaries, which accompany the Stanzas.
>
> In the apocalyptic ślokas of the Archaic Records, the language is as symbolical, if less mythical, than in the Purānas. Without the help of the later *commentaries,* compiled by generations of adepts, it would be impossible to understand the meaning correctly. In the ancient Cosmogonies, the visible and the invisible worlds are the double links of one and the same chain. As the invisible *Logos,* with its seven hierarchies (represented or personified each by its chief angel or rector), form one POWER, the inner and the invisible; so, in the world of Forms, the Sun and the seven chief Planets constitute the visible and active potency; the latter "Hierarchy" being, so to speak, the visible and objective *Logos* of the invisible and (except in the lowest grades) ever-subjective angels.[13]

The Significance of the Logos

Logos is often used as an equivalent term for a Regent. However, a qualifying adjective is required in order to designate the status or grade of the Logos. Thus in the quotation reference was made to the invisible Logos (which is usually termed the First Logos, or the Unmanifest Logos) and further to the objective Logos, signifying the Third Logos or the Creative Logos. In Brahmanical philosophy these are referred to as Brahman and Brahmā. Philosophically, Logos is a Greek term signifying the cosmic reason giving order and purpose to the world, even though the literal meaning is a "word." Hence it is often rendered into Latin by means

---

13. Ibid. II:23; III:36, 6-vol. ed.; II:26-7, 3rd ed.

of the term Verbum, which also means a "word." When the idea of pronouncing a word *(logos)* is formulated in the mind, it may be said that this stage represents the unmanifest aspect of the word—the First Logos. Then the idea of the word must be conveyed to the appropriate organs of speech for the production of the word. This represents the concept of the Second Logos—which may be said to represent the bridge linking the unmanifest to the stage of manifestation. Hence the Second Logos is often regarded in a "hyphenated aspect" as the Unmanifest-manifest Logos. Although the organs of speech are not actually visible, they are made use of for the next stage, which is the actual utterance of the word by means of the creative ability exercized by the mind. Thus when the word is uttered this represents the Manifested or Third Logos, also termed the Creative Logos. The functioning of the creative aspect is inherent in the First Logos and is conveyed by means of the Second Logos to the Third Logos, thereby becoming manifest.

> The esoteric meaning of the word *Logos* (speech or word, *Verbum*) is the rendering in objective expression, as in a photograph of the concealed thought. The Logos is the mirror reflecting DIVINE MIND, and the Universe is the mirror of the Logos, though the latter is the *esse* of that Universe. As the *Logos* reflects *all* in the Universe of Pleroma, so man reflects in himself all that he sees and finds in his Universe, the Earth . . . . "Every Universe (world or planet) has its own Logos," says the doctrine. The Sun was always called by the Egyptians "the eye of Osiris," and was himself the *Logos,* the first-begotten, or light made manifest to the world, "which is the Mind and divine intellect of the Concealed." It is only by the sevenfold Ray of this light that we can become cognizant of the Logos through the Demiurge, regarding the latter as the *creator* of our planet and every thing pertaining to it, and the former as the guiding Force of that "Creator."[14]

---

14. Ibid. II:25; III:38, 6-vol. ed.; II: 28-9, 3rd ed. Observe that the Egyptians called the Sun "the eye of Osiris," just as other ancient peoples referred to the Sun as "the eye of the Lord," as mentioned previously. In ancient Persia, the Sun symbolized "the eye of Ormuzd."

The Seven Regents

In further reference to the sun's giving life to the seven, a supplementary paragraph may now be included:

> "The seven higher make the Seven Lhas create the world," states a Commentary; which means that our Earth, leaving aside the rest, was *created* or fashioned by terrestrial spirits, the "Regents" being simply the supervisors. This is the first germ, the seed of that which grew later into the Tree of Astrology and Astrolatry. The Higher ones were the *Kosmocratores,* the fabricators of our solar system. This is borne out by all the ancient Cosmogonies: that of Hermes, of the Chaldees, of the Aryans, of the Egyptians, and even of the Jews. Heaven's belt, the signs of the Zodiac (the *Sacred animals*), are as much the B'ne Alhim (Sons of the Gods or the Elohim) as the Spirits of the Earth; but they are prior to them. Soma and Sin, Isis and Diana, are all lunar gods or goddesses, called the fathers and mothers of our Earth, which is subordinate to them. But these, in their turn, are subordinate to their "Fathers" and "Mothers"—the latter interchangeable and varying with each nation—the gods and their planets, such as Jupiter, Saturn, Bel, Brihaspati, etc.[15]

The Kosmocratores denote Regents of a superior degree, for the word itself means Builders of the cosmos, applicable to a solar system or even a planetary system. They represent the personification of the Creative Forces. These divine Builders "build on planes in the Divine Mind, visible to them though not to us."[16]

> Damascius repeats the teachings of the Pagan writers when he explains that "There are seven series of cosmocratores or cosmic forces, which are double: the higher ones commissioned to support and guide the superior world; the lower ones, the inferior world our own."
>
> And he is but saying what the ancients taught. Iamblichus gives this dogma of the duality of all the planets

15. Ibid. II:23; III:36, 6-vol. ed.; II:26, 3rd ed.

16. Ibid. III:561; V:534, 6-vol. ed.

> and celestial bodies, of gods and daimons (spirits). He also divides the Archontes into two classes—the more and the less spiritual; the latter more connected with and clothed with matter, as having a *form*, while the former are bodiless (*arūpa*).[17]

The Archontes are usually regarded as "rulers," but here they are equivalent to the Regents.

Soma is one of the Sanskrit names associated with the Moon, so is Sin in Chaldea, and Isis in Egypt. Diana was the ancient Roman name for the goddess of the Moon; in Greece the lunar goddess was known as Artemis. Bel is the Assyrian and Babylonian deity equivalent to Jupiter or Zeus; while Brihaspati is the Sanskrit name for the planet Jupiter.

There is yet one more sentence in the first śloka which continues the topic of the sun as the giver of life. Even though brief it may be elucidated by means of the Esoteric Philosophy: "It gave life to the first." The sun is referred to by the word "it." "The first" may be applied in several ways: (1) to the first humanity which came to the Earth during the First Round; (2) to the first land in the Fourth Round (this first land is regarded as the First Continent, Ādi-Varsha); (3) to the first planetary manifestation of the system which is now represented by the Earth-Chain. Four more meanings of "the first" are available, with appropriate explanations:

(4) The first Globe of the planetary chain of the Earth (Globe A):

> The Globe, propelled onward by the Spirit of the Earth and his six assistants, gets all its vital forces, life, and powers through the medium of the seven planetary Dhyānis from the Spirit of the Sun. They are his messengers of Light and Life.[18]

The six assistants are the six other globes of the Earth-Chain. The seven planetary Dhyānis signify the Regents of the Seven Sacred Planets.

---

17. Ibid. III:321; V:315, 6-vol. ed.

18. Ibid. II:28-9; III:41, 6-vol. ed.; II:32, 3rd ed.

(5) The first of the seven regions of the Earth in the Fourth Round: because each one of the seven regions of the Earth

> . . . receives its light and life from its own especial Dhyāni—spiritually, and from the palace (house, the planet) of that Dhyāni physically.[19]

(6) The First Race in the Fourth Round: because

> . . . every Race in its evolution is said to be born under the direct influence of one of the Planets: Race the first receiving its breath of life from the Sun.[20]

(7) The first of the Seven Sacred Planets:

> Every active power or force of the Earth comes to her from one of the seven Lords.[21]

That is to say, each one of the globes of the Earth-Chain gets its vitality through each one of the Seven Sacred Planets. The seven Lords signify the seven Regents of the Seven Sacred Planets.

Further, in regard to the Regents, or Watchers, there is

> . . . one placed as the Guardian or Regent over each of the seven divisions of the Earth.[22]
>
> As our Sun radiates its light and beams on every body in space within the boundaries of its system, so the Regent of every Planet-star, the Parent-monad, shoots out from itself the Monad of every "pilgrim" Soul born under its house within its own group. The Regents are esoterically seven, whether in the Sephiroth, the "Angels of the Presence," the Rishis, or the Amshaspends.[23]

---

19. Ibid. II:29; III:41, 6-vol. ed.; II:32, 3rd ed.

20. Ibid. II:23-4; III:37, 6-vol. ed.; II:27, 3rd ed.

21. Ibid. II:29; III:41, 6-vol. ed.; II:32-3, 3rd ed.

22. Ibid. III:369; V:356, 6-vol. ed.

23. Ibid. III:341; V:333, 6-vol. ed.

5.

THE SEVEN SACRED PLANETS

IT MAY appear strange to have a chapter on the Seven Sacred Planets in a book on the peopling of the Earth, but the following exposition will show the reason for the association as well as its preeminent significance.

Everyone knows that there are more than seven planets in the solar system, so there must be a purpose for calling only seven of them "sacred." Yet when the heavenly bodies which revolve around the sun are named, only five—Mercury, Venus, Mars, Jupiter, Saturn—are included in the seven "sacred planets." The listing is completed by adding the Sun and the Moon, but these may be regarded as substitutes for planets which are not at present visible.

The Seven Planets of the Ancients

H. P. Blavatsky's commentary on why the seven planets were regarded as sacred is excellently stated:

> The seven planets are not limited to this number because the ancients knew of no others, but simply because they were primitive or primordial *houses* of the seven Logoi. There may be nine and ninety-nine other planets discovered—this does not alter the fact of these seven alone being sacred.[1]

Along with the idea that the sacred planets were represented as the "houses" of the seven Logoi, it was brought out in the chapter on the Planetary Regents[2] that the seven

1. SD II:602 fn.; IV:173, fn. 6-vol. ed.; II:637, fn. 3rd ed.

2. Chapter 4, "The Planetary Regents."

sacred planets were regarded as the houses or mansions of the Watcher of the planet—one Watcher for each planet. The connotation "sacred" also signified that one of the seven Logoi of the seven sacred planets acted as a Regent for one of the seven globes of the Earth planetary system during its coming into being as well as during its existence.

How the ancients regarded the seven planets is explained:

> All the classical authors and philosophers who have treated the subject repeat, with Hermes Trismegistus, that the seven Rectors—the planets including the sun—were the associates, or the co-workers, of the Unknown All represented by the Demiourgos, commissioned to contain the Cosmos—our planetary world—within seven circles. Plutarch shows them representing "the circle of the celestial worlds." Again, Denys of Thracia and the learned Clemens of Alexandria both describe the Rectors as being shown in the Egyptian temples in the shape of mysterious wheels or spheres always in motion, which made the Initiates affirm that the problem of perpetual motion had been solved by the celestial wheels in the Initiation Adyta.[3]

One of the easiest ways of remembering the seven sacred planets is by correlating them with the days of the week, most of which are actually named after the ancient Scandinavian deities.

Days of the Week	*Scandinavian Deity*	*Latin name of Planet*
Sunday	Sunnan	Sol (substitute)
Monday	Monan	Luna (substitute)
Tuesday	Tiw (god of War)	Mars
Wednesday	Woden or Odin	Mercury
Thursday	Thor	Jupiter
Friday	Friga or Freya	Venus
Saturday	Saeter or Saetern	Saturn

In the same way the seven Angels or seven Spirits of the Presence, highly regarded in Roman Catholic literature, may

3. SD III:328-9; V:321, 6-vol. ed.

be made equivalent to the Regents of the planets: Sun—Michael; Moon—Gabriel; Mars—Samael; Mercury—Raphael; Jupiter—Zachariel; Venus—Anael; Saturn—Orifiel.

Origen, the Christian Father, mentioned that the Gnostics made this listing of the Seven Sacred Planets: Adonai, equivalent to the genius of the Sun; Iao, of the Moon; Eloi, of Jupiter; Sabao, of Mars; Orai, of Venus; Astaphai, of Mercury; Ildabaoth, of Saturn.[4]

The Messengers of Light and Life

Another reason why the Logoi or Watchers were held in such high esteem is that the seven planetary Dhyānis were regarded as the Messengers of Light and Life from the sun; for they act as transmitters of the vital forces, life, and powers to the Earth. The One Life-Force, which is the Life-Essence sustaining suns and planets, is not located in the solar orb but is focused through and by means of the sun from the inner realms of being.

> . . . the origin of the LIFE ESSENCE, Occultism locates in the same center as the nucleus of prima materia (for they are one) of our Solar system.
>
> The Sun is the heart of the Solar World (System) and its brain is hidden behind the (visible) Sun. From thence, sensation is radiated into every nerve-center of the great body, and the waves of the life-essence flow into each artery and vein. . . . The planets are its limbs and pulses. (Commentary)
>
> It was stated elsewhere . . . that Occult philosophy denies that the Sun is a globe in combustion, but defines it simply as a world, a glowing sphere, the *real* Sun being hidden behind, and the visible being only its reflection, its *shell*. The Nasmyth willow leaves, mistaken by Sir W. Herschel for "Solar inhabitants," are the reservoirs of solar vital energy, "the vital electricity that feeds the whole system. . . . The Sun *in abscondito* being thus the storehouse of our little Kosmos, self-generating its vital fluid, and ever receiving as much as it gives out," and the *visible* Sun only a *window cut into*

4. Origen, *Contra Celsum*, Bk. VI, xxiv-xxxviii.

> *the real* Solar palace and presence, which reflects, however, faithfully the interior work.
>
> Thus, there is a regular circulation of the vital fluid throughout our system, of which the Sun is the heart—the same as the circulation of the blood in the human body—during the manvantaric solar period, or life; the Sun contracting as rhythmically, at every return of it, as the human heart does. Only, instead of performing the round in a second or so, it takes the solar blood ten of its years, and a whole year to pass through its *auricles* and *ventricles* before it washes the *lungs* and passes thence to the great veins and arteries of the system.[5]

The planets are regarded as the limbs and pulses, for they act as the means of circulating the life-essence from planet to planet and on to the sun. In this manner the life-essence is circulated to each member of the system. At the same time each globe of the planetary system is sustained by means of the life-essence, which is conveyed to each portion of a globe throughout the system. It is in this way that the Seven Sacred Planets act as the sun's "Messengers of light and life."

Just as there is a Lha who is the Regent of the Earth, so there are six other Regents, one for each of the other six globes of the planetary system of Earth. All seven Regents are termed Dhyāni-Bodhisattvas; they are the superior beings who watch over or have regency of the globes of the system. But in a still loftier stage of evolutionary attainment are the seven planetary Dhyānis who act as intermediaries for passing on to the Earth-Chain the vital forces, life, and powers issuing from the Spirit of the Sun—or the Solar Logos; these are the planetary Logoi of the Seven Sacred Planets. They are also referred to as the "Seven Sons of Light."

H. P. Blavatsky's method of naming the Seven Sacred Planets was certainly a singular one, and it will be noted that neither Uranus, Neptune, nor Pluto was included in the enumeration:

> It is then the "Seven Sons of Light"—called after their planets and (by the rabble) often identified with

5. SD I:540-1; II:264-5, 6-vol. ed.; I:590-1, 3rd ed.

> them—namely Saturn, Jupiter, Mercury, Mars, Venus, and—*presumably* for the modern critic, who goes no deeper than the surface of old religions—the Sun and Moon, which are, according to the Occult teachings, our heavenly Parents, or "Father," synthetically. . . .[6]

Further, in regard to the Seven Sacred Planets, these are

> . . . the heavenly bodies in direct astral and psychic communication with the Earth, its Guides, and Watchers—morally and physically; the visible orbs furnishing our Humanity with its outward and inward characteristics, and their "Regents" or *Rectors* with our Monads and spiritual faculties.[7]

It will be observed that the Sun and the moon were not regarded as "planets," but as "heavenly parents"; in fact, a note was added stating that the sun and moon were used "as substitutes for esoteric purposes."[8]

The Chaldeans designated Seven Sacred Planets, and the same names were used by the Assyrians and Babylonians:

Shamash	—	Sun (substituting for a planet)
Marduk	—	Jupiter
Ishtar	—	Venus
Ninib	—	Saturn
Nebo	—	Mercury
Nergal	—	Mars
Sin	—	Moon (substituting for a planet)

The names of the Seven Sacred Planets in Sanskrit are defined in a commentary from the Book of Dzyan, which indicates also that each of the seven Rectors has regency over one of the seven primordial human groups, in connection with the peopling of the earth. Furthermore, the regency is extended to the Root-Races.

> Like each of the seven regions of the Earth, each of the seven First-born (the primordial human groups)

---

6. Ibid. I:575; II 298-9, 6-vol. ed.; I:628, 3rd ed.

7. Ibid.

8. Ibid.

> receives its light and life from its own especial Dhyāni—spiritually, and from the palace (house, the planet) of that Dhyāni—physically; so with the seven great Races to be born on it. The first is born under the Sun; the second under Brihaspati (Jupiter); the third under Lohitānga (the "fiery-bodied," Venus, or Śukra); the fourth, under Soma (the Moon, our Globe also, the Fourth Sphere being born under and from the Moon) and Śani, Saturn, the Krūra-lochana (evil-eyed) and the Asita (the dark); the fifth, under Budha (Mercury).[9]

H. P. Blavatsky remarks that this enumeration of the regency of the Dhyānis—which applies to the seven regions of the earth, to the seven principles of man, as well as to the Root-Races—is applicable also to "seven states of matter and the seven forms of force, [and] the division must be the same in all that concerns the earth."[10] Therefore it would apply to the seven globes of the Earth-Chain. The regency of each of the seven globes is now indicated, by listing the Sanskrit names along with the corresponding familiar names of the planets, together with their astronomical symbols:

Sūrya	Sun (substituting for a planet)	⊙	Globe A
Brihaspati	Jupiter	♃	Globe B
Uśanas-Śukra (or Lohitānga)	Venus	♀	Globe C
Śani (or Krūra-lochana)	Saturn	♄	Globe D
Budha	Mercury	☿	Globe E
Angāraka	Mars	♂	Globe F
Soma	Moon (substituting for a planet)	☽	Globe G

A diagram may be used to represent the position held by each of the Seven Sons of Light over one of the seven globes of the Earth planetary system. This will serve a fourfold

9. Ibid. II:29; III:41-2, 6-vol. ed.; II:32, 3rd ed.

10. Ibid.

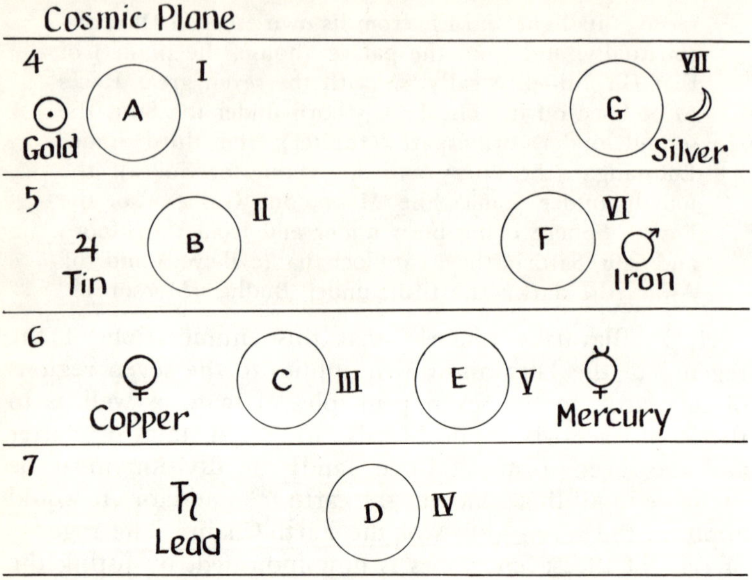

DIAGRAM 5

In this diagram the four lower planes, termed "manifested," are depicted and enumerated as Cosmic Planes 4-7. The seven globes forming the Earth planetary system are positioned on these planes in the following manner: On Cosmic Plane 4, Globes A and G; on Cosmic Plane 5, Globes B and F; on Cosmic Plane 6, Globes C and E; on Cosmic Plane 7, Globe D. Alongside each one of the seven globes are placed the astronomical symbols for the Seven Sacred Planets. This indicates the planetary regency of each one of the seven globes of the Earth system. The Roman numeral placed alongside of each one of the seven globes indicates the Root-Race over which a particular Sacred Planet has regency.

purpose: (1) to illustrate the position of the seven globes in regard to four of the seven cosmic planes; (2) to depict the regency of the Dhyānis over specific globes of the Earth system by use of astronomical symbols; (3) to show in Roman numerals the supervision over particular Root-Races; and (4) to indicate the metal associated with the specific planets.

Each one of the Seven Sacred Planets is of course a septenary global system: this was emphasized by H. P. Blavatsky:

> The *seven* planets are the sacred planets of antiquity, and are all septenary.[11]

11. Ibid. I:167; I:221, 6-vol. ed.; I:191, 3rd ed.

The Plea of the Lha of the Earth

The Stanzas of Dzyan contain a direct reference to the peopling of the earth in the first sentence of the second śloka of the first stanza. The description which has been provided concerning the Seven Sacred Planets is a necessary stage in the portrayal of this process, showing as it does the kinship of the planets to the Earth system. This is referred to in the Stanzas. Replacing the narrative style of the opening stanza, the second śloka recounts the words addressed to the Solar Logos by the Dhyāni of the Earth:

> Said the Earth:—"Lord of the Shining Face; my house is empty. . . . send thy sons to people this wheel. Thou hast sent thy seven sons to the Lord of Wisdom. Seven times doth he see thee nearer to himself, seven times more doth he feel thee. Thou hast forbidden thy servants, the small rings, to catch thy light and heat, thy great bounty to intercept on its passage. Send now to thy servant the same." (Stanza I, śloka 2)[12]

The Planetary Spirit, or Lha, of the fourth globe of the Earth system calls upon the Lord of the Shining Face, the great Chohan of the solar system. The significant point about the "emptiness of the house" is that it refers to the long period during which the globe has been left without the human Life-Wave—a term which signifies the evolutionary aspect of the human kingdom as it passes from globe to globe of the Earth system.

It should be borne in mind that the human Life-Wave manifests on the Earth-Chain for the purpose of accomplishing seven evolutionary circuits (termed Rounds). A circuit is accomplished by means of evolutionary stages on each one of the seven globes of the Earth-Chain in successive order. Seven major evolutionary stages (termed Root-Races) are accomplished on each of the seven globes. During the third principal circuit (or Round), the human Life-Wave had indeed accomplished seven major developmental stages of evolution on Globe D (usually referred to as the seven

---

12. Ibid. II:15; III:27, 6-vol. ed.; II:15, 3rd ed.

Root-Races—sometimes called "rings"). Yet these evolutionary stages constituted only a small portion of the requisite circuit comprising the Third Round. Following this accomplishment, the human Life-Wave was then obliged to continue the circuit, or life-cycle, on the Ascending Arc, thus leaving the "house empty of evolutionary human activity." The Ascending Arc comprises seven major evolutionary stages on each of the three superior globes, Globes E, F, and G. Upon completion of the Ascending Arc of the third circuit, after an appropriate resting period, the human Life-Wave began the cycle of the Descending Arc on the next circuit, termed the Fourth Round. This consisted of seven major evolutionary stages on the three first globes of the series: on Globe A, descending to Globe B, then a further descent to Globe C. Thus while the Ascending and Descending Arcs, or cycles, were being accomplished, Globe D was regarded as being "empty." Consequently, the Regent of the Earth is portrayed as requesting the great Lord of the Shining Face to send his sons to people the fourth globe—which is referred to as a "wheel." One interpretation of "the Sons" would be the seven classes of monads who people a planet.

The śloka continues: "Thou hast sent thy seven sons to the Lord of Wisdom." The name of the Lord of Wisdom is the Regent of the planet Mercury: in Sanskrit "Budha," which means wisdom. Just as the planetary system of Earth has seven globes, in like manner the planetary system of Mercury has its seven globes, as does every other planet. This is referred to in the śloka by the words "the seven sons": one Regent for each one of the globes of the Mercury planetary system. The next portion of the sentence—"seven times doth he see thee nearer to himself"—is explained in *The Secret Doctrine* to mean that "Mercury receives seven times more light and heat from the Sun than Earth."[13]

The significance of the concluding sentences of the śloka: "Thou hast forbidden thy servants, the small rings, to catch thy light and heat, thy great bounty to intercept on its pas-

---

13. Ibid. II 27-8; III:40, 6-vol. ed.; II:31, 3rd ed.

sage. Send now to thy servant the same," (śloka 2) was clarified in the quotation from the Commentary from the Book of Dzyan, when reference was made to the fact that the earth "gets all its vital forces, life, and powers through the medium of the seven planetary Dhyānis from the Spirit of the Sun." This is further emphasized in this passage from the Commentary:

> Every active power or force of the earth comes to her from one of the seven Lords.[14]

The response of the Lord of the Shining Face to the Dhyāni's request to people his wheel is the subject of the next chapter.

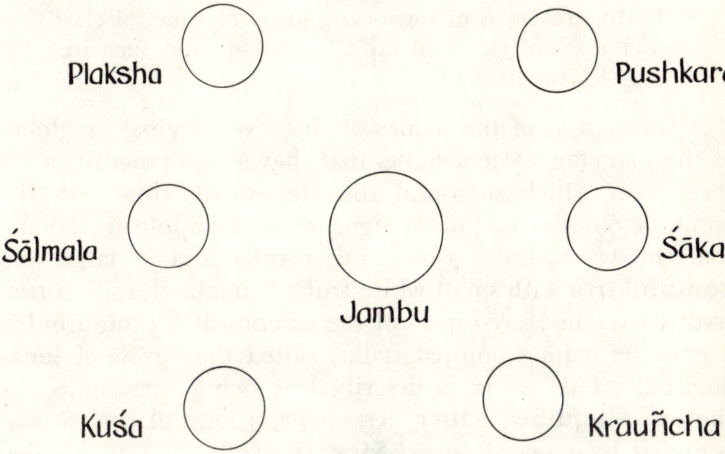

DIAGRAM 6

Additional confirmation of the existence of seven worlds which compose a planetary system is available in the Purānas. But the passage in which they are considered is related in an allegorical manner, so that the true significance of the seven worlds is certainly not made apparent. In the first place they are referred to in connection with legendary characters as well as fanciful names, consequently instead

14. Ibid. II:29; III:42, 6-vol. ed.; II:32-3, 3rd ed.

of worlds they are termed Dvīpas (a word signifying "islands") thus concealing the real meaning. In similar vein *The Secret Doctrine* thus alludes to the Purānic account of the seven worlds:

> The Seven *Dvīpas* apportioned to Priyavrata's septenary progeny refer to several localities: first of all to our planetary chain. *Jambu-dvīpa* alone representing our globe, the six others are the (to us) invisible companion globes of this earth. This is shown by the very nature of the allegorical and symbolic descriptions. *Jambu* (dvīpa) *"is in the center of all these* [the so-called insular continents] and is surrounded" by a *sea of salt water* (lavaṇa), whereas Plaksha, Śalmala, Kuśa, Krauñcha, Śāka and Pushkara, are "surrounded severally–by great seas of sugar-cane juice, of wine, of clarified butter, of curds, of milk," etc., etc., and such like metaphorical names.[15]

A translation of the names of the seven Dvīpas or globes of the planetary system shows that they are as fanciful as are the "seas" which surround the globes. Starting with the name of our Earth, Jambu signifies a rose-apple tree; Plaksha, the waved-leaf fig-tree, Ficus Iṇfectoria, a large and beautiful tree with small white fruit; Śalmala, the silk-cotton tree; Kuśa, the sacred grass of the ascetics, Poa cynosuroides, a grass with long pointed stalks, called the "grass of lucky augury." This world is described as being surrounded by the sea of liquified butter; Krauñcha, a kind of curlew, surrounded by a sea of curds; Śāka, the teak tree, surrounded by a sea of milk or white sea; Pushkara, the blue lotus.

---

15. *Vishnu-Purāna* (Wilson's translation), Book II:ii, quoted in SD II:320-1; III 320, 6-vol. ed.; II:334, 3rd ed.

6.

THE LUNAR PITRIS

WHO ARE the beings who brought about the peopling of the Earth? The Stanzas of Dzyan provide this information in the reply given to the Planetary Spirit of the Earth by the Solar Logos: "Thy people shall be under the rule of the Fathers." (śloka 3) The Fathers are known as the Pitris, the Sanskrit word meaning fathers. Inasmuch as the Dhyāni (or Regent) of the Earth had appealed to the Lord of the Shining Face for assistance in peopling the Earth, the Great Lord responded: "Apply to thy father, the Lord of the Lotus, for his sons." The Lord of the Lotus is none other than the Dhyāni of the Moon, consequently when appearing on Earth the Sons of the Lord of the Lotus are known as the Lunar Pitris.

But these are not the only names given to the Lunar Pitris. They are in fact called by so many different names in *The Secret Doctrine* that listing them will be of assistance.

Fathers	Sons of the Lord of the Lotus
Ancestors	Sons of Soma
Celestial Ancestors	Celestial Men
Progenitors	Pitris
Divine Progenitors	Pitaras
Lunar Progenitors	Lunar Pitris
Lunar Ancestors	Pitar-devatās
Lunar Fathers	Barhishads
Lunar Monads	Barhishad Pitris
Lunar Spirits	Corporeal Pitris
Lunar Gods	Dhyānis
Moon-Gods	Lords of the Moon
Lunar Sons	Will-born Lords
Sons of the Moon	Sons of Twilight

Descriptive Names of the Pitris

Each name has a distinctive connotation. Definitions will indicate the role of the Pitris in the peopling of the Earth.

FATHERS. The idea of parentage is applied in a general way to the beings who brought about the peopling of the Earth.

ANCESTORS. Here there is an association with a line of descent more ancient in character than is suggested by "Fathers."

CELESTIAL ANCESTORS. In addition to the idea of parentage, there is now an indication that this parentage is of a superior, heavenly status. The word "celestial" is used in *The Secret Doctrine* for the beings who are superior to humans, that is superior in their stage of evolutionary development, having already evolved through the human stage in previous manvantaras (or cyclical periods of evolutional activity).

PROGENITORS. This term has associations with begetting and the birth aspect of the parental function.

DIVINE PROGENITORS. While continuing with the birth aspect, the relationship to divinity is here established.

LUNAR ANCESTORS. This term emphasizes our descent from moon beings.

LUNAR PROGENITORS. Again the specific source of our ancestry is stressed.

LUNAR FATHERS. The immediate relationship of the Pitris ("Fathers") to the humans on Earth is indicated by this term.

LUNAR MONADS. The Seven-Round cycle of evolutionary development (termed a manvantara) having been accomplished in its cyclical progression on the moon, it is now being continued on Earth in the kingdoms of Nature. The human stage of evolution on Earth continues the development of monads who had just arrived at the human stage on the moon.

MONADIC EVOLUTION. The monad uses the vestures (or rūpas) of the kingdoms of Nature for a manvantaric evolutionary experience, passing from one kingdom to another at the end of each manvantara.

LUNAR SPIRITS. This has nearly the same significance as Lunar Monads, but it indicates the immortal or spiritual aspects of the former Moon beings.

LUNAR GODS. Emphasis is placed on the evolutionary attainment which was accomplished by the beings who, on the Moon, were in a position equivalent to our present human stage of evolutionary development on the Earth.

MOON-GODS. An alternate term for Lunar Gods.

LUNAR SONS. The beings who were undergoing evolutionary development on the Moon were in their turn assisted by superior beings; hence the humans on the Moon were also "sons" of superior beings.

SONS OF THE MOON. An alternate term for Lunar Sons.

SONS OF THE LORD OF THE LOTUS. In the Stanzas of Dzyan the Lord of the Lotus is mentioned as the Father of the Planetary Spirit of the Earth, consequently the relationship of the Earth and the Moon are made apparent when reference is made to the Lord of the Lotus. In Sanskrit, Kumuda-pati signifies the Lord of the Lotus; therefore his Sons are named Kumuda-putras.

SONS OF SOMA. An alternate term for Lunar Sons. Soma is one of the Sanskrit words used in connection with the Moon, especially in the Commentaries on the Stanzas of Dzyan.

CELESTIAL MEN. Although it might seem illogical to state that there are "celestial" men, in view of the fact that the term "celestial" is usually associated with heavenly beings (such as deities or angels rather than humans), nevertheless it is significant that the human beings on the Lunar Chain who had successfully accomplished the Seven-Round cycle on that series of seven globes, had actually graduated from the human kingdom. However, because the Lunar Chain is inferior to the Earth Chain, these Lunar Monads had not yet attained sufficient evolutionary status to enter and function in the kingdom superior to the human kingdom on the Earth Chain. The superior kingdom is termed the Dhyāni-Chohanic Kingdom. Therefore, the designation of Celestial Men is significantly applicable to the Sons of the Lord of the Lotus.

PITRIS. This is the Sanskrit word of which the literal meaning is "Fathers." The word is derived from the verbal root *pā,* meaning to watch, guard, protect, shelter, rule, govern.

PITARAS. The nominative plural of Pitri: one of the eight cases in which a noun is declined in Sanskrit grammar.

LUNAR PITRIS. Attention is here focused on the Lunar aspect of the Fathers, in contradistinction to the *Solar* Pitris.

PITAR-DEVATĀS. *Devatā* signifies the quality of being divine. The compound word stresses the divine quality of the Pitris and is an allusion to the fact that the Pitris are in a category superior to the human kingdom. *Pitar* is the vocative case of *pitri.*

BARHISHADS. This Sanskrit word defines the status of the Pitris in a particular way. The word is compounded of *barhiś,* "sacred grass" or "fire" and *sad,* to sit; hence "those who sit by the fire." Although this is a very literal derivation, it provides a clue to understanding the role that the Pitris play in the peopling of the earth. "Fire" is a symbol of man's fifth principle, Manas, the mind principle. Because the Lunar Pitris in their evolutionary development had not yet attained the status of being able to awaken the "fire of mind" in nascent humanity on the Earth, they are represented (by the word "Barhishad") as "sitting by the fire." The *Solar* Pitris are the awakeners of the "fire of mind"—the mind principle.

BARHISHAD PITRIS. Literally "the Fathers who sit by the fire."

> Barhishad Pitris, or the Pitar-Devatās, i.e., those *possessed of the physical creative fire.* They could only create, or rather clothe, the human Monads with their own astral Selves, but they could not make man in their image and likeness.[1]

CORPOREAL PITRIS. This term emphasizes a specific function of the Lunar Fathers: that of producing the form for "the first Manushyas on this Earth," that is, the first men. For of the seven classes of Pitris, "these Beings fall into four

1. SD, II:94; III:103, 6-vol. ed.; II:99, 3rd ed.

corporeal (or grosser) and three incorporeal (or subtler) 'principles.' "[2] The "four corporeal" signify the Lunar Pitris; the "three incorporeal" are named the Solar Pitris—those who awaken the "fire of mind" in humanity.

DHYĀNIS. A Sanskrit word used as an abbreviation for Dhyāni-Chohans. The word *dhyāni* is an adjective utilized in compound forms (in place of dhyānin). As stated earlier, the noun, *dhyāna,* signifies meditation. The compound, Dhyāni-Chohan, is a generalizing term for any being superior in status to that of the human kingdom. Care must be taken to distinguish between a Lunar Dhyāni, a Solar Dhyāni, or a Dhyāni of the Seven Sacred Planets—Dhyāni here denoting a Regent.

LORDS OF THE MOON. This is one of the names by which the Lunar Pitris are described in the Stanzas of Dzyan, specifically in Stanza III.

THE WILL-BORN LORDS. This indicates the prime function of the Lunar Pitris: producing forms for the monads waiting for incarnation into the human kingdom on Earth (the fourth globe) during the Fourth Round. In Stanza IV the Will-born Lords are described as "separating men from themselves, . . . propelled by the spirit of Life-giving."

SONS OF TWILIGHT. Another reference to the Lunar Pitris, in Stanza VI. The significance of the term is best explained by one of the mythological stories told in the *Purānas* about Brahmā, the Creative Logos (or Third Logos).

Brahmā concentrated his mind and assumed another form—one in which the quality of great passivity prevailed. From his side there then issued the Pitris, the progenitors of men. This body of Brahmā, the "Body of Twilight," when cast off, became the Sandhyā, the evening twilight, the interval between day and night. Hence the Pitris are called Sons of Twilight. Men were produced from Brahmā in this body of Day. The significance of the two "bodies of Brahmā," Day and Twilight, is this: the "Body of Day" refers to the present planetary system, the Earth; the "Body of Twilight" refers to the Moon-Chain, whence the pitris came—from a previous

2. Ibid. II:92; III:101, 6-vol. ed.; II:96, 3rd ed.

planetary system—the Twilight which preceded that of "the Day."

How the Lunar Pitris provided the forms (or rūpas) for the peopling of the Earth will be told in a subsequent śloka in the Stanzas.

The Response to the Plea of the Lha of the Earth

In the second śloka of the first Stanza, the Lha of the Earth, the Planetary Dhyāni or Logos, called upon the Lord of the Shining Face (the Sun) for assistance. The third śloka opens with the Solar Lord's reply.

> Said the "Lord of the Shining Face": "I shall send thee a fire when thy work is commenced. Raise thy voice to other Lokas; apply to thy father, the Lord of the Lotus, for his sons. . . . Thy people shall be under the rule of the Fathers. Thy men shall be mortals. The men of the Lord of Wisdom, not the Lunar Sons, are immortal. Cease thy complaints."[3]

Obviously the words in the first sentence of the Great Lord's response are symbolic. Thus the use of the word "fire" is significant. Since the previous cycles—specifically three Circuits, or three Rounds—had not brought forth "the fire," very likely "the fire of mind" is referred to, rather than the cosmic Element-Principle of Fire. Mention of the cosmic Element-Principle of Fire refers to the teaching of the Esoteric Philosophy that during every Circuit (or Round) of the Earth system one cosmic Element-Principle is brought into manifestation on the Earth-Chain and undergoes a sevenfold evolutionary development. Thus during the first Circuit (or First Round) the cosmic Element-Principle of Fire, named Tejas-Tattva, was made manifest on the Earth. The cosmic Element-Principle at present manifesting on this globe and undergoing an evolutionary development during this Fourth Round is termed Prithivī-Tattva (the Element-Principle of Earth). However, the "fire of mind," referred to here, came into manifestation during a later cycle (or

3. Ibid. II:15; III:56, 6-vol. ed.; II:48, 3rd ed.

age) than the period described in the first sentence of the third śloka.

Since the Lord of the Shining Face declared that he would send "a fire when thy work is commenced," it is evident that the necessary "work" of preparation for the peopling of the Earth had not yet started (this will be explained in the next chapter), although a period of dormancy obtained and a further waiting period ensued. When a period of evolutionary activity of the human kingdom is concluded on a globe of the planetary system, a period of dormancy (which has been termed an "obscuration") follows. The period of activity comprises seven major developmental stages of evolution. Thus the requisite "work" on the Earth had not yet commenced; Globe D of the Earth-Chain was not yet ready; in fact it was in a state of dormancy. Nevertheless, the seven cycles of evolutionary development during the period of activity on Globe C had been concluded. This is borne out by the words of the Planetary Spirit of the Earth: "My house is empty. Send thy sons to people this wheel."

The "waiting period" or preparatory period at the conclusion of seven major developmental stages of evolution on a globe of the planetary system is a resting period, termed a *Sandhyā.* This Sanskrit word means "the junction between night and day." It is derived from the verb-root *samdha,* to join, to unite, usually defined as a "twilight." This resting period is equivalent to one-tenth of the duration of the period of activity just concluded, comprising seven stages of evolutionary development. In this case it represents the period of activity concluded on Globe C of the Earth-Chain. When the Sandhyā is terminated, there is an equal similar period of rest termed a *Sandhyānśa* (dawn) *before* the next period of activity dawns.

"Raise thy voice to other Lokas; apply to thy father, the Lord of the Lotus, for his sons," continued the Solar Lord, in the second sentence of the śloka. *Lokas* here refer to the superior planes, where the Spirit of the Earth's father, Kumuda-Pati, is stationed. *Kumuda* is one of the names of the lotus; *pati* signifies lord. Lord of the Lotus is a name applied to the Lord of the Moon, who is described as dwelling in

Soma-loka (the "dwelling of the Moon"). Here he is designated "thy father." This clearly indicates that the Moon is regarded as the parent of the Earth. The supreme hierarch of a planetary system, termed the hyparxis, is situated on the highest plane of the system, represented here by the words "other lokas." The Earth, Globe D, is situated on the lowest plane, the seventh cosmic plane (counting downward).

The Men of the Earth Will Be Mortals

Since the Solar Lord advised the Planetary Dhyāni to apply to his father, the Lord of the Lotus, "for his sons," the beings who will assist in the peopling of the Earth *will come from the Moon.* They are Pitris who have accomplished the Seven-Round cycle of evolution on the Moon.

A break in the response of the Solar Lord is indicated by an ellipsis. Then the Lord of the Sun continues: "Thy people shall be under the rule of the Fathers." (This sentence has been fully considered.) After the word "Fathers," *The Secret Doctrine* places the term *Pitri-pati,* a Sanskrit compound signifying the Lord of the Pitris. An important observation may be made here in connection with Pitri-pati. As well as meaning Lord of the Moon, it is one of the names of Yama, the god of death, he who judges the dead in Yama-loka, the Underworld. The Solar Lord had declared: "Thy people shall be under the rule of the Fathers. Thy men shall be mortals;" indicating that the "men of Earth" are to be subject to death, for they come under the rule of the god of death, Yama. "The men of the Lord of Wisdom not the sons of Soma are immortal," stated the Solar Lord. The Lord of Wisdom is Budha, the planetary system of Mercury; the Lord of Soma is the Moon, the planetary system or chain of which the present Moon is the representative on the seventh cosmic plane. It is significant that "the men of Mercury" are referred to as "immortal." This would seem to imply that there *are* beings on Mercury who have passed through the stage that the "men of Earth" are now passing through, although the human beings of Earth are at the present time far lower in the stage of evolutionary development than the "men of Mercury," who are here described as "immortal." Evidently they are not burdened with a physical vehicle which by the law of cycles, of coming and going, must be subject to death.

In connection with the mortality of the terrestrial men *The Secret Doctrine* adds:

> The Moon being an inferior body to the Earth even, to say nothing of other planets, the terrestrial men produced by her sons—the lunar men or "ancestors"—from her shell or body, cannot be immortal. They cannot hope to become real, self-conscious and intelligent men, unless they are finished, so to say, by other creators.[4]

The "finishing" process was indeed carried out "by other creators," namely the Solar Pitris. (See Chapter 13.)

---

4. Ibid. II:45; III:56, 6-vol. ed.; II:48, 3rd ed.

7.

PREPARATORY STAGES FOR THE PEOPLING OF THE EARTH

In the first chapter the planetary system of the Earth was described as being composed of seven globes. The concept was presented as a postulate reading: " . . . every sidereal body, every planet, whether visible or invisible, is credited with six companion globes."[1] A second postulate presents the idea that our Earth was not always in its present state of materiality or physicality, because it also experiences differing stages of evolution, just as it is requisite for all the kingdoms of nature to pass through different vehicular changes in conformity with the changes that occur to the Earth during the cycles of the Rounds. Here is the postulate:

> Our Earth, as the visible representative of its invisible superior globes, its "lords" or "principles," has to live, as have the others, through seven Rounds. During the first three, it forms and consolidates; during the fourth it settles and hardens; during the last three it gradually returns to its first ethereal form: it is spiritualized, so to say.[2]

---

1. SD, I:158; I:213, 6-vol. ed.; I:182, 3rd ed.

2. Ibid. I:159; I:213, 6-vol. ed.; I:182-3, 3rd ed.

The Earth Undergoes Changes

A Commentary from the Book of Dzyan describes the changes to which the Earth is subjected:

> Archaic Scripture teaches that at the commencement of every local Kalpa, or Round, the earth is reborn; "as the human Jīva (Monad), when passing into a new womb, gets re-covered with a new body, so does the Jīva of the Earth; it gets a more perfect and solid covering with each Round after re-emerging once more from the matrix of space into objectivity." This process is attended, of course, by the throes of the new birth or geological convulsions.[3]

In the above quotation the word Jīva is equivalent to the term "Monad" and this is the case in several passages in the Commentaries. Jīva does not here refer to the cosmic principle of Life. The Sanskrit word *Jīva* is derived from the verb root *jīv,* meaning to live; hence its literal significance is *life,* especially in its relation with the One Life or Life-Force: that which is all-pervasive during the period of activity of a planetary system. When this One Life-Force becomes individualized in connection with a human being it is termed Prāna, one of the principles of the sevenfold constitution of man.

The reference to the Earth getting "a more perfect and solid covering with each Round" is also applicable to the human rūpa or form in its Round evolution. However, during the present stage of human evolutionary development the rūpa is not as dense or as solid as it was during its most material phase, which was attained at the lowest point of the Descending Arc. This condition occurred at the middle point of the Fourth Round.

The words "matrix of space" from which the Jīva of the Earth emerges, present a concept which is probably better understood by the term Laya Center, which is defined in this manner: a cosmos comes into manifestation because its "inner principles" (those which are undying and thus en-

---

3. Ibid. II:46; III:58, 6-vol. ed.; II:50, 3rd ed.

able the cosmos to reemerge from a dormant center) have been quiescent in a laya center, namely the point of matter where every differentiation has ceased. This is also referred to as a zero point, indicating a point at which the scale of reckoning of differentiation begins. A laya center may also be regarded as the disappearing point for all things below it and the entering point for all things above it.

When the process of the departure of life from a planet takes place, resulting in what is known as the moon-stage, this is due to the transfer of the "inner principles" of such a planet into a laya center.

> It is Fohat who guides the transfer of the principles from one planet to the other, from one star to another—child-star. When a planet dies, its informing principles are transferred to a *laya* or sleeping center, with potential but latent energy in it, which is thus awakened into life and begins to form itself into a new sidereal body.[4]

It should be remembered that each one of the seven globes of the planetary system forms its own individual laya center, out of which each one of the seven globes emerges.

> When Fohat is said to produce "Seven Laya Centers," it means that for formative or creative purposes, the GREAT LAW (Theists may call it God) stops, or rather modifies its perpetual motion on seven invisible points within the area of the manifested Universe.[5]

The "Coverings" of the Earth

The Commentary from the Book of Dzyan defines progressive stages which the Earth undergoes. It also states that at certain cyclic periods changes took place in its "coverings." In the Stanzas of Dzyan these "coverings" are referred to as "skins" of the Earth, and during each cyclic period (which is termed a Round) the Earth is said to put on "seven skins." The seven skins represent seven evolutionary stages of development.

---

4. Ibid. I:147; I:202, 6-vol. ed.; I:170, 3rd ed.

5. Ibid. I:147; I:203, 6-vol. ed.; I:171, 3rd ed.

Before it was possible for the beginning of the peopling of the Earth to take place, certain preparatory stages were requisite. The concluding words of the third śloka refer to this in three brief sentences. Doubtless because of their brevity their significance is apt to be overlooked. These concise statements are spoken by the Lord of the Shining Face (the Sun) to the Planetary Spirit of the Earth:

> "Thy seven skins are yet on thee. . . . Thou are not ready. Thy men are not ready." (Stanza I, śloka 3)[6]

The Commentary from the Book of Dzyan (previously quoted) states that at every Round the Earth will get "a more perfect and solid covering." This "covering" consists of "seven skins." The clue to the meaning of the words spoken by the Solar Lord is provided in the fourth śloka of Stanza I, where it is told that the Earth "casts off her old three." This signifies that the Earth has passed through three Circuits (or Rounds) and is preparing for the fourth Circuit. Nevertheless, the śloka declares that its "seven skins" are yet on the Earth. These "seven skins" signify the seven major developmental stages which are accomplished during one cyclic period of activity (termed one Round) on each one of the globes of the planetary system. The fact that *three* Rounds had been accomplished is declared in the fourth śloka. The Earth is there described as "casting off her old three skins."

The second sentence of the third śloka, "Thou art not ready," means that the Earth is not fully prepared for the peopling to take place. It further indicates that at this period the "old three skins" had not yet been cast off. In fact the fourth śloka tells how this is done.

The concluding sentence of the third śloka affirms that the "men are not ready," that is to say, the monads are not yet ready because they have not yet awakened from the period of dormancy which prevails during the sandhyā. This interval (the sandhyā) takes place between the exit

---

6. Ibid. II:15; III:27, 6-vol. ed.; II:16, 3rd ed.

from one globe and the entrance to another globe of the system.

The narrative form which began the account is resumed in the fourth śloka, replacing the terse dialogue of the second and third ślokas.

> After great throes she cast off her old three and put on her new seven skins, and stood in her first one.[7]

Although a clarification of the seven skins has been given, there is apt to be some confusion between the old three, the new seven, and the Earth's first one. The key to understanding this portion of the stanza is given by means of the adjectives *old* and *new*. Then, too, as with so many of the ślokas and their commentaries, there are *two sets of sevens*. The first set of sevens signifies the Rounds; the second set stands for the seven major developmental stages of evolution, termed the seven Root-Races of the human kingdom's evolutionary developmental stages. This term is certainly not suitably applied to the stages of development of the Earth.

Thus, after great throes the Earth "cast off her *old three*," signifying the three Rounds of the seven Cycles, or Circuits, which the Earth "has to live through," according to the fundamental postulate. Putting on "her *new seven skins*" refers to the seven *new* major developmental stages of evolution which the Earth will undergo during the Fourth Round. Standing in "her *first one*" represents the *first* of the seven major developmental stages of evolution which must be accomplished during the Fourth Round. Each time a new skin is put on, one at a time, a new stage of evolution is attained.

The Secret Doctrine explains śloka 4:

> The Earth is said to cast off her old *three* skins, because this refers to the three preceding Rounds she has already passed through; the present being the *fourth* Round out of the seven. At the beginning of every new

7. Ibid. II:15; III:28, 6-vol. ed.; II:16, 3rd ed.

> Round, after a period of "obscuration,"[8] the earth (as do also the other six earths) casts off, or is supposed to cast off, her old skins as the Serpent does: therefore she is called in the *Aitareya-Brāhmana* the *Sarpa-Rājñī*, "the Queen of the Serpents," and "the mother of all that moves." The "Seven Skins," in the first of which she now stands, refer to seven geological changes which accompany and correspond to the evolution of the Seven Root Races of Humanity.[9]

The Seven Geological Changes

As there will be seven stages of evolutionary development on the Earth, equivalent to seven Root-Races, there will be seven geological changes. The human kingdom is now undergoing its fifth major evolutionary developmental stage, or evolving in the fifth Root-Race, and there have been five major geological changes. These changes of the Earth's surface are associated with the rise and fall or the formation of continents, signifying the inclusion of all the land-masses on the globe which exist at any time during the time-period of a Root-Race.

The first continent is named the Imperishable Sacred Land—Ādi-varsha. It is so called

> . . . because it is the only one whose destiny it is to last from the beginning to the end of the Manvantara throughout each Round. . . According to a poetical expression in one of the Commentaries, the "pole-star has its watchful eye upon it, from the dawn to the close of the twilight of 'a day' of the Great Breath."[10]

This "Day" is a Day of Brahmā, equivalent to seven Rounds.

---

8. The "period of obscuration" is that interval which has been previously described as the period of dormancy which occurs when the host of human monads (or the human Life-Wave) has left one globe of the Earth-Chain in order to enter the next of the system's globes in serial order. The passage through all seven globes of the Earth system is one Round, or one Circuit of the seven globes.

9. SD II:47; III:58-9, 6-vol. ed.; II:50-1, 3rd ed.

10. Ibid. II:6; III:19-20, 6-vol. ed.; II:6, 3rd ed.

The second Continent was called the Hyperborean.

> Its promontories extended southward and westward from the North Pole to receive the Second [Root-] Race, and comprised the whole of what is now known as Northern Asia. . . . It was a real continent, a *bona fide* land which knew no winter in those early days.[11]

The name of Lemuria was given to the third Continent, in place of its real name.

> It not only embraced a vast area in the Pacific and Indian Oceans, but extended in the shape of a horseshoe past Madagascar, round "South Africa" (then a mere fragment in process of formation), through the Atlantic up to Norway.[12]

The above description had reference to Northern Lemuria. As for Southern Lemuria:

> It covered the whole area of space from the foot of the Himalayas, which separated it from the inland sea rolling its waves over what is now Tibet, Mongolia, and the great desert of Shamo (Gobi); from Chittagong westward to Hardvar, and eastward to Assam. From thence, it stretched South across what is known to us as Southern India, Ceylon, and Sumatra; then embracing on its way, as we go South, Madagascar on its right hand and Australia and Tasmania on its left, it ran down to within a few degrees of the Antarctic Circle; when, from Australia, an inland region on the Mother Continent in those ages, it extended far into the Pacific Ocean, not only beyond Rapa-nui (Teapy, or Easter Island) which now lies in latitude 26 S., and longitude 110 W.[13]

Atlantis was the name of the Fourth Continent.

> The *Atlantic portion of Lemuria* was the geological basis of what is generally known as Atlantis. The latter, indeed, must be regarded rather as a development of

---

11. Ibid. II:7; III:20, 6-vol. ed.; II:6-7, 3rd ed.

12. Ibid. II:333; III:332-3, 6-vol. ed.; II:348, 3rd ed.

13. Ibid. II:323-4; III:323-4, 6-vol. ed.; II:338, 3rd ed.

the Atlantic prolongation of Lemuria, than as an entirely new mass of land upheaved to meet the special requirements of the Fourth Root-Race. . . . Their continent was formed by the coalescence of many islands and peninsulas which were upheaved in the ordinary course of time *and became ultimately the true home of the great Race known as the Atlanteans.*[14]

. . . at different epochs of the past huge islands, and even continents existed where now there is but a wild waste of waters . . . at a remote epoch a traveller could traverse what is now the Atlantic Ocean, almost the entire distance by land, crossing in boats from one island to another, where narrow straits then existed.[15]

Some words on the overlapping of the continents and their sinking and upheaval may now be added:

In the Eocene Age—even in its "very first part," the great cycle of the fourth Race men, the Atlanteans—had already reached its highest point, and the great continent, the father of nearly all the present continents—showed the first symptoms of sinking—a process that occupied it down to 11,446 years ago, when its last island, that, translating its vernacular name, we may call with propriety *Poseidonis*—went down with a crash. By the bye, *whoever* wrote the Review of Donnelly's *Atlantis* is right: Lemuria can no more be confounded with the Atlantic Continent than Europe with America. Both sank and were drowned with their high civilizations and "gods," yet between the two catastrophes, a short period of about 700,000 years elapsed; "Lemuria" flourishing and ending her career just at about that trifling lapse of time before the early part of the Eocene Age, since its race was the *third*. . . Why not admit—true no one of them has ever thought of it—that our *present* continents, have—like "Lemuria" and "Atlantis"—been *several times already,* submerged and had the time to reappear again, and bear their new groups of mankind and civilization; and that, at the first great geological

14. Ibid. II:333-4; III:333, 6-vol. ed.; II:348, 3rd ed.

15. *Isis Unveiled,* I:558.

upheaval, at the next cataclysm—in the series of periodical cataclysms that occur from the beginning to the end of every Round—our already *autopsized* continents will go down, and the Lemurias and Atlantises come up again.[16]

The Fifth Continent was America; but, as it is situated at the Antipodes, it is Europe and Asia Minor, almost coeval with it, which are generally referred to by the Indo-Aryan Occultists as the fifth. If their teaching followed the appearance of the Continents in their geological and geographical order, then this classification would have to be altered. But as the sequence of the Continents is made to follow the order of evolution of the Races, from the first to the fifth, our Aryan Root-race, Europe must be called the fifth great Continent. The Secret Doctrine takes no account of islands and peninsulas, nor does it follow the modern geographical distribution of land and sea. Since the day of its earliest teachings and the destruction of the great Atlantis, the face of the earth has changed more than once. There was a time when the delta of Egypt and Northern Africa belonged to Europe, before the formation of the Straits of Gibraltar, and a further upheaval of the continent, changed entirely the face of the map of Europe. The last serious change occurred some 12,000 years ago, and was followed by the submersion of Plato's little Atlantic island, which he calls Atlantis after its parent continent.[17]

Here is a forecast of some future geological changes:

Of course your Science is right in many of her generalities, but her premises are wrong, or at any rate—very faulty. For instance she is right in saying that while the new America was forming the ancient Atlantis was sinking, and gradually washing away; but she is neither right in her given epochs nor in the calculations of the duration of that sinking. The latter—is the future fate of your British Islands the first on the list of victims that have to be destroyed by fire (submarine volcanos) and

16. ML, p. 151; pp. 147-8, 3rd ed.

17. SD II:8-9; III:21-2, 6-vol. ed.; II:8, 3rd ed.

water, France and other lands will follow suit. When they reappear again, the last seventh Sub-race of the sixth Root race of present mankind will be flourishing on "Lemuria" and "Atlantis," both of which will have reappeared also (their reappearance following immediately the disappearance of the present isles and continents), and very few seas and *great waters* will be found then on our globe, waters as well as land appearing and disappearing and shifting periodically and each in turn.[18]

---

18. ML p. 156; pp. 152-3, 3rd ed.

8.

THE STATE OF THE EARTH PRIOR TO ITS PEOPLING

THE ARCHAIC manuscripts which relate the account of the peopling of the Earth provide the only source of information about the stages of the world prior to the advent of man. A reference to the earliest state of the Earth is given in a Commentary from the Book of Dzyan. It tells how the Earth planetary system comes into manifestation from a state of nonmanifestation.

> After the changeless (avikāra) immutable nature (Essence, sadaikarūpa) had awakened and changed (differentiated) into (a state of) causality (avyakta), and from cause (Kārana) had become its own discrete effect (vyakta), from invisible it became visible. The smallest of the small (the most atomic of atoms, or anīyāmsam aniyāsam) became one and the many (ekānekarūpa); and producing the Universe produced also the Fourth Loka (our Earth) in the garland of the seven lotuses. The Achyuta then became the Chyuta.[1]

The Sanskrit words are placed after their English equivalents for clarification. A further exposition may be helpful.

When a planetary system enters into the state of pralaya, a period of complete rest from all evolutionary processes prevails, for pralaya signifies a dissolution of all forms. Yet the system's essential substance remains unchanged during the period. The Sanskrit word *avikāra* means unchanged.

---

1. SD, II:46-7; III:68, 6-vol. ed.; II:50, 3rd ed.

The significance of the dissolution of forms may be illustrated by placing lumps of sugar in very hot water. The sugar quickly dissolves and permeates the liquid; the lumps (forms) are in a state of dissolution.

The Awakening of a Planetary System

When the time for awakening of a planetary system arrives, Fohat—the universal, basic energy—becomes more active and arouses the sleeping globes; in fact Fohat provides the stimulus for the awakening. The essential substance, which had remained quiescent during the period of the pralaya, becomes reactivated; it enters into a state of causality, and this in turn becomes the manifested, or the invisible becomes the visible.

Although the cause of the coming into manifestation of a planetary system is described as a cosmic atom (the most atomic of atoms—the smallest of the small), it has tremendous potency—sufficient to produce a world. On a still vaster scale the power is there to produce a whole universe. Thus from the one come the many. The cosmic atom produces the Earth planetary system, described in the Commentary as a garland of seven lotuses, signifying the seven globes of the system. Our Earth is mentioned as the Fourth Loka, that is the fourth world of the septenary chain of globes.

After the closing sentence of the quotation, namely: "The Achyuta then became the Chyuta," this footnote was added:

> Achyuta is an almost untranslatable term. It means that which is not subject to fall or change for the worse: the Unfalling; and it is the reverse of chyuta, "the Fallen."[2]

This may be applied to the birth of a world system in this manner: that which had existed as the imperishable (also the permanent) now manifests as the perishable, the impermanent; for achyuta also signifies imperishable, permanent; chyuta, perishable and impermanent.

---

2. Ibid. II:47; III:58, 6-vol. ed.; II:50, 3rd ed.

Another Commentary describes the emergence from the Laya state—the state where substance becomes homogeneous and is unable to act or differentiate:

> Whatsoever quits the Laya State, becomes active life; it is drawn into the vortex of MOTION (the alchemical solvent of Life); Spirit and Matter are the two States of the ONE, which is neither Spirit nor Matter, both being the absolute life, latent. (Book of Dzyan, Commentary III, par. 18)[3]

Obviously the Earth and its elements in their primary stages were not in the same condition or state in which they are now:

> The elements, whether simple or compound, could not have remained the same since the commencement of the evolution of our chain. Everything in the Universe progresses steadily in the Great Cycle, while incessantly going up and down in the smaller cycles. Nature is never stationary during manvantara, as it is ever *becoming,* not simply *being*; and mineral, vegetable, and human life are always adapting their organisms to the then reigning Elements, and therefore *those* Elements were then fitted for them, as they are now for the life of present humanity.[4]

With regard to the subject of the elements and their relationship in the formation of the Earth, the Commentary explains:

> The worlds, to the profane are built up of the known Elements. To the conception of an Arhat, these Elements are themselves collectively a divine Life; distributively, on the plane of manifestations, the numberless and countless crores of lives. Fire alone is ONE, on the plane of the One Reality: on that of manifested, hence illusive, being, its particles are fiery lives which live and have their being at the expense of every other life that they consume. Therefore they are named the "DEVOURERS." . . . Every visible thing in this universe

3. Ibid. I:258; I:302, 6-vol. ed.; I:278-9, 3rd ed.

4. Ibid. I:257; I:301, 6-vol. ed.; I:278, 3rd ed.

was built by such Lives, from conscious and divine primordial man down to the unconscious agents that construct matter.[5]

The Earth is built up for the first Round by the "Devourers" which disintegrate and differentiate the germs of other lives in the Elements.[6]

The Condition of the Earth in the Early Rounds

When the "Devourers" have differentiated "the fire-atoms" by a peculiar process of segmentation, the latter become life-germs, which aggregate according to the laws of cohesion and affinity. Then the life-germs produce lives of another kind, which work on the structure of our globes.

Thus, in the first Round, the globe, having been built by the primitive fire-lives, i.e., formed into a sphere—had no solidity, nor qualifications, save a cold brightness, nor form nor color; it is only towards the end of the First Round that it developed one Element which from its inorganic, so to say, or simple Essence became now in our Round the fire we know throughout the system. The Earth was in her first rūpa.[7]

The Earth, such as we know it now, had no existence before the 4th Round, hundreds of millions of years ago, the commencement of our geological Earth. The globe was *"fiery, cool and radiant* as its ethereal men and animals during the first Round," says the Commentary. . . . *"luminous* and more dense and heavy—during the second Round."[8]

The Commentary contains another brief reference to the Second Round of the Earth, signifying its second evolutionary phase of development:

From the second Round, Earth—hitherto a foetus in the matrix of Space—began its real existence: it had

5. Ibid. I:249-50; I:294, 6-vol. ed.; I:269-70, 3rd ed. (1 crore is equivalent to ten million years.)

6. Ibid. I:258; I:302, 6-vol. ed.; I:278, 3rd ed.

7. Ibid. I:259; I:303, 6-vol. ed.; I:279, 3rd ed.

8. Ibid. I:252 fn.; I:297, 6-vol. ed; I:273, 3rd ed.

developed individual sentient life, its second principle.[9]

Still less information is given concerning the condition of the Earth during the Third Round—its third phase of evolutionary development. A Commentary described it simply as "watery" (but not "water" as we now know it). Nevertheless, the production of forms for the lower kingdoms is mentioned in connection with the preparatory stages for the peopling of the Earth. This is related in the fifth śloka (of the Stanzas of Dzyan). Prior to this, a Commentary is quoted regarding cataclysms which took place upon the Earth:

> During the first seven crores of the Kalpa (70,000,000 years) the Earth and its two Kingdoms (mineral and vegetable), one already having achieved its seventh circle, the other, hardly nascent, are luminous and semi-ethereal, cold, lifeless, and translucid. In the eleventh crore the mother (Earth) grows opaque, and in the fourteenth the throes of adolescence take place. These convulsions of nature (geological changes) last till her twentieth crore of years, uninterruptedly, after which they become periodical, and at long intervals.
>
> The last change took place nearly twelve crores of years ago (120,000,000). But the Earth with everything on her face had become cool, hard and settled ages earlier. (Commentary, xxii)
>
> Thus, if we are to believe esoteric teaching, there have been no more *universal* geological disturbances and changes for the last 120 millions of years, and the Earth was, even before that time, ready to receive her human stock.[10]

In the above quotation, the figures which amount to seven crores and are placed after the Sanskrit word *Kalpa*, are not applicable to the time-period of a Kalpa. The latter word usually stands for a Day of Brahmā—which signifies the life-period of a planetary Manvantara and consists of

---

9. Ibid. I:260; I:304, 6-vol. ed.; I:280, 3rd ed. The significance of "the matrix of space" was defined in Chapter 7.

10. Ibid. II:312; III:312-3, 6-vol. ed.; II:326, 3rd ed.

seven Rounds or 4,320,000,000 years. Reference to the achievement of the "seventh circle" signifies that seven evolutionary stages were accomplished by the Mineral Kingdom in the First Round; whereas the Plant Kingdom was "hardly nascent." One crore is equivalent to ten million years.

The Earth During a Period of Dormancy

The early evolutionary stages of the Earth are narrated in the Stanzas of Dzyan. The fifth śloka, which opens the second Stanza, continues the theme of the fourth śloka. It refers to the state of the Earth during a period of dormancy—which means the period when the Human Kingdom had not yet begun to execute its evolutionary cycles on the lowest globe of the Earth Chain.

> The Wheel whirled for thirty crores (of years, or 300,000,000—or Three Occult Ages). It constructed rūpas (forms). Soft stones, that hardened (minerals); hard plants, that softened (vegetation). Visible, from invisible, insects and small lives (sarīsripa, svapada). She (the Earth) shook them off her back, whenever they overran the mother. (śloka 5).
>
> This relates to an inclination of the axis—of which there were several—to a consequent deluge and chaos on Earth (having, however, no reference to primeval chaos.)[11]

The "Wheel" here is the Earth (Globe D of the planetary system). The two Sanskrit words are now explained: *sarīsripa*—creeping or crawling animals; *svapada*—literally "self-going."

With regard to the first manifestation of plants, *The Secret Doctrine* adds an explanatory passage which follows after a quotation from the second chapter of Genesis:

> "... the Lord God made the earth and the heavens, and every plant of the field before it was in the earth, and

11. Ibid. II:52; III:63, 6-vol. ed.; II:55, 3rd ed.

> every herb of the field before it grew: for the Lord God had not caused it to rain upon the earth" (Ch. 2:4-5) —an absurdity unless the esoteric explanation is accepted. The plants *were* created before they were in the earth—*for there was no earth then such as it is now*; and the herb of the field was in existence before it grew as it does now in the fourth Round.[12]

An explanation is also provided as to how the forms of minerals, plants and even animals, which were quite different in preceding Rounds, came into manifestation on this Round.

> The astral prototypes of the mineral, vegetable and animal kingdoms up to man have taken that time (300 million years) to evolve, re-forming out of the cast-off materials of the preceding Round [the Third Round], which, though very dense and physical in their own cycle, are relatively ethereal as compared with the materiality of our present middle Round. At the expiration of these 300 million years, Nature, on the way to the physical and material, down the arc of descent, begins with mankind and works downwards, hardening or materializing forms as it proceeds. Thus the fossils found in strata, to which an antiquity, not of eighteen, but of many hundreds of millions of years, must be ascribed, belong in reality to forms of the preceding Round, which, while living, were far more ethereal than physical, as *we know the physical.* That we perceive and disinter them as tangible forms, is due to the process of materialization or crystallization referred to, which took place subsequently, at the beginning of the Fourth Round, and reached its maximum after the appearance of man, proceeding parallel with his physical evolution. This alone illustrates the fact that the degree of materiality of the Earth changes *pari passu* with that of its inhabitants. And thus man now finds, as tangible fossils, what were once the (to his present senses) ethereal forms of the lower kingdoms.[13]

---

12. Ibid. I:254; I:298, 6-vol. ed.; I:274, 3rd ed.

13. Ibid. II 68 fn.; III:78 fn., 6-vol. ed.; II:72, 3rd ed.

With reference to the early stones and plants of this earth, a passage explains that these were the results of previous evolutionary development: they were produced in the earlier Round—the Third Round:

> . . . these stones, plants, and animals were the prototypes, the filmy presentments of those of the Fourth Round; and even those at the beginning of the Fourth Round were the astral shadows of the present, as the Occultists express it. And finally the forms and *genera* of neither man, animal, nor plant were what they became later. Thus the astral prototypes of the lower beings of the animal kingdom of the Fourth Round, which *preceded* the chhāyās of *Men,* were the consolidated, though still very ethereal *sheaths* of the still more ethereal forms or models produced at the close of the Third Round on Globe D. "Produced from the residue of the substance matter; from dead bodies of men and (other extinct) animals of the wheel before," or the previous *Third Round*—as śloka 28 tells us.[14]

14. Ibid. II:186; III:193, 6-vol. ed.; II:196-7, 3rd ed. The significance of the term *Chhāyās* and how they were produced is explained in Chapter 12.

9.

"NATURE UNAIDED FAILS"

THIS is the heading which introduces Stanza II of the second series of the Stanzas of Dzyan. The narrative style continues but the subject, portraying events of long bygone eras, is entirely different. The account reads like a myth; but it is said that history is often preserved in legends and myths. The significance of the second stanza is that there is supervision over the kingdoms of nature during the periods of evolutionary activity as well as during periods of rest or dormancy. When a kingdom is executing its seven stages of evolutionary development, superior Beings of the Dhyâni-Chohanic Kingdoms are guiding or watching over the process. "Nature *is* aided" is a statement of truth.

The Stanzas of Dzyan recount the aid that has been provided to the human kingdom in each of its stages of development, of which each Root-Race is a major stage. In the First Root-Race the Lunar Pitris gave their help; in the Second, aid came from the Asuras; in the Third and Fourth Root-Races the four classes of the Solar Pitris—named the Mānasaputras, Agnishvāttas, Kumāras, Vairājas—provided their assistance. Then the Fifth Root-Race "was ruled over by the first divine Kings" in its early subraces.

If some higher level of control is not exercised during periods of dormancy, inharmonious results occur; consequently "Nature unaided fails." In fact, Stanza II tells of such an occurrence, for it describes a period of dormancy during the lowest stage of the Earth's history: an era in the Arc of Descent– the "descent into matter." It is during the midway Round of the midway globe that the element-principle of earth, known as the Prithivītattva is in process of evolution.[1] The midway Round of the seven-Round Circuit is equivalent to three and one-half Rounds; the midway globe is the fourth of the septenary series.

The Creation of Forms Without Guidance

The first portion of śloka 5, which begins Stanza II and which considers the early stages of the Earth, has already been dealt with in Chapter 8. The first portion of the śloka mentions a time-period of thirty crores, signifying three Occult Ages, or 300 million years. The second portion also refers to the three Occult Ages:

> After thirty crores of years she [the Earth] turned round. She laid on her back; on her side. . . . She would call no sons of Heaven, she would ask no sons of Wisdom. She created from her own bosom. She evolved water-men, terrible and bad.[1]

The Secret Doctrine explains:

> This relates to an inclination of the axis–of which there were several–to a consequent deluge and chaos on Earth (having, however, no reference to primeval chaos), in which monsters, half-human, half-animal, were generated. We find it mentioned in the "Book of the Dead," and also in the Chaldean account of creation, on the Cutha Tablets, however mutilated.
>
> It is not even allegory. Here we have *facts,* that are found repeated in the account of *Pymander,* as well as in the Chaldean tablets of creation. The verses may almost be checked by the Cosmogony, as given by Berosus, which has been disfigured out of recognition by Eusebius, but some of the features of which may yet be found in fragments left by ancient Greek authors–Apollodorus, Alexander Polyhistor, etc., etc.[2]

---

1. This will be explained further on in the present chapter.

2. SD II:52-3; III:63, 6-vol. ed.; II:55-6, 3rd ed.

Berosus was a priest of Bel at Babylon, who lived about 300 B. C. He compiled the history of his native land from documents, which he translated into Greek, and also included in his three books the Babylonian knowledge of astronomy and astrology. Although his works were published in 250 B.C. during the reign of Antiochus, they have not come down to our day. However, extracts from his history were made by Josephus, the Jewish historian, as well as by Eusebius, bishop of Caesarea, the ecclesiastical historian. Berosus added Babylonian mythology to his historical account. In this portion of his work he told of Oannes, describing him as a god, part man and part fish, a Piscean deity, who resided in the Persian Gulf.

Eusebius Pamphili—better known as Eusebius of Caesarea, because he was the bishop of Caesarea in Palestine—is regarded as the outstanding ecclesiastical historian who lived during the years of 260 to 340 A.D. In addition to his historical writings, he is famed for his apologetic works of the early church, of which he wrote more than fifteen books. Some of his dogmatic and polemical writings also have come down to our day.

Apollodorus, born about 180 B.C., was an Athenian grammarian, a pupil of Aristarchus and the Stoic philosopher, Panaetius. After leaving Alexandria for Pergamum about 146 B.C. he settled in Athens, where he did most of his writing. He is best known for his *Chronicle,* highly regarded as Greek comedy. Other works by him included a treatise *On the Gods,* the Homeric *Catalogue of the Ships,* as well as his critical grammarian writings.

Alexander Cornelius, born at Miletus, although a Greek grammarian, became a Roman citizen in 82 B.C., after being taken to Rome as a prisoner in the Mithridatic war. Because of his great learning and his historical writings he was surnamed Polyhistor.

The Pymander is a Hermetic work, known as *The Divine Pymander,* meaning divine thought. In regard to this work H. P. Blavatsky wrote:

> How truly esoteric and consonant with the Secret Doctrine is "Pymander the Thought Divine" of Hermes, may

be inferred from its original and primitive translations in Latin and Greek only.[3]

The closing sentence of śloka 5 relates what happened when creation of forms was attempted without the guidance of the superior Beings: "She the Earth evolved water-men, terrible and bad." *The Secret Doctrine* comments:

> "The water-men terrible and bad," who were the production of physical nature alone, a result of the "evolutionary impulse" and the first attempt to create *man* the "crown," and the aim and goal of all animal life on Earth—are shown to be failures in our Stanzas. Do we not find the same in the Berosian Cosmogony, denounced with such vehemence as the culmination of heathen absurdity? And yet who of the Evolutionists can say that things in the beginning have not come to pass as they are described? That, as maintained in the Purānas, the Egyptian and Chaldean fragments, and even in Genesis, there have not been two, and even more, "creations" before the last formation of the Globe; which, changing its geological and atmospheric conditions, changed also its flora, its fauna, and its men?[4]
>
> The Chaldean fragments of Cosmogony on the Cuneiform inscriptions, and elsewhere, show two distinct creations of animals and men, the first being destroyed, as it was a failure.[5]
>
> Oannes (or Dagon, the Chaldean "Man-fish") divides his Cosmogony and Genesis into two portions. First the abyss of waters and darkness, wherein resided most hideous beings—men with wings, four and two-faced men, human beings with two heads, with the legs and horns of a goat (our "goat-men"), hippocentaurs, bulls with the heads of men, and dogs with tails of fishes. In short, combinations of various animals and men, of fishes, reptiles and other monstrous animals assuming each other's shapes and countenances.[6]

---

3. Ibid. II:491; IV:59, 6-vol. ed.; II:516, 3rd ed.
4. Ibid. II:53; III:63-4, 6-vol. ed.; II:56, 3rd ed.
5. Ibid. II:54; III:64, 6-vol. ed.; II:57, 3rd ed.
6. Ibid. II:54; III:64, 6-vol. ed.; II:57, 3rd ed.

A passage in *The Secret Doctrine,* deals with the production of what are termed "nondescript animals," and explains that:

> . . . the astral prototypes of the lower beings of the animal kingdom of the Fourth Round, which *preceded* . . . *Men,* were the consolidated, though still very ethereal *sheaths* of the still more ethereal forms or models produced at the close of the Third Round on Globe D. "Produced from the residue of the substance matter; from dead bodies of men and (other extinct) animals of the wheel before," or the previous Third Round—as Śloka 28 tells us.[7]

How the Lower Kingdoms Were Created

Śloka 6 continues the description of what occurred in the early time-periods as presented in ślokas 4 and 5. This is referred to by the words "first, second and third," which signify the evolutionary cycles of the three previous Rounds. The śloka clearly shows that divine assistance is on hand to determine whether or not the results of nature's unaided efforts are suitable for the monads of the human kingdom. It should be borne in mind that the human Life-Wave had not yet come upon the Earth nor had it started its evolutionary development on the fourth globe (Globe D, our Earth), but was still on the third globe, Globe C. The śloka reiterates the result of the failure of unaided nature to provide fit rūpas (forms), and also explains how this took place:

> The water-men terrible and bad she herself created. From the remains of others, from the dross and slime of her first, second, and third, she formed them.[8]

The Cosmic Element-Principles

After the words "remains of others" in the above citation, H. P. Blavatsky added: "from the mineral, vegetable and

7. Ibid. II:186; III:193, 6-vol. ed.; II:197, 3rd ed.

8. Ibid. II:16; III:28, 6-vol. ed.; II:16, 3rd ed.

animal remains." The word "Rounds" was placed after the "first, second and third." The significance of the evolutionary development which occurs during the three Rounds preceding the Fourth is best explained by the successive development of the cosmic Element-Principles.[9] For *The Secret Doctrine* postulates that during every Round, one of the seven Tattvas, or Cosmic Element-Principles, is brought into manifestation and proceeds with its evolutionary processes. Each Element-Principle undergoes its specific evolutionary stages.

> This means that every new Round develops one of the Compound Elements, as now known to Science, which rejects the primitive nomenclature, preferring to subdivide them into constituents. If Nature is the "Ever-becoming" on the manifested plane, then those Elements are to be regarded in the same light: they have to evolve, progress, and increase to the Manvantaric end. Thus the First Round, we are taught, developed but one Element, and a nature and humanity in what may be called one aspect of Nature—called by some, very unscientifically, though it may be so *de facto,* "One-dimensional Space."[10]

The first Cosmic Element-Principle, brought forth and developed during the First Round, was Fire. In Sanskrit it is named Tejas-tattva. The condition of the Earth in the First Round is also described:

> Thus, in the first Round, the globe, having been buiit by the primitive fire-lives—i.e., formed into a sphere—had no solidity, nor qualifications, save a cold brightness, nor form nor color; it was only toward the end of the First Round that it developed one Element which from its inorganic, so to say, or simple Essence became now in our Round the fire we know throughout the sys-

9. The term "Element-Principle" is used as an equivalent to the Sanskrit word *Tattva,* so as not to confuse the word "element" with the hundred or so elements of modern science. The "four elements" of the ancients—Fire, Air, Water, Earth—are to be viewed as the proper or natural environment and constitution of something (as explained in the dictionary under the word "element").

10. SD I:250; I:295, 6-vol. ed.; I:269-70, 3rd ed.

> tem. The Earth was in her first rūpa, the essence of which is the Ākāśic principle . . . now known as, and very erroneously termed, Astral Light.[11]

During the Second Round the second cosmic Element-Principle of Air, Vāyu-tattva, was emanated:

> The Second Round brings into manifestation the second element—Air, that element, the purity of which would ensure continuous life to him who would use it.[12]

Following the advent of this second Element-Principle, the third Element-Principle, Water or Apas-tattva, appeared in the Third Round. It was during the process of the changes which occurred in the Fourth Round—but while the Water-Principle was still predominant—that the production of the "water-men" occurred. Thus they were formed from "the remains" of the Third Round evolution. It was during the Fourth Round that the cosmic Element-Principle of Earth, named Prithivī-tattva, came into manifestation.

> The Third Round developed the *third* principle—*Water*; while the Fourth transformed the gaseous fluids and plastic form of our globe into the hard, crusted, grossly material sphere we are living on. "Bhūmi" has reached her fourth principle.[13]

Bhūmi signifies our earth, applicable to the fourth globe, Globe D.

> The succession of primary aspects of Nature with which the succession of Rounds is concerned, has to do, as already indicated, with the development of the "Elements" (in the Occult sense)—Fire, Air, Water, Earth. We are only in the fourth Round, and our catalog so far stops short. The centers of consciousness (destined to develop into humanity as we know it) of the third Round arrived at a perception of the third Element, Water.

---

11. Ibid. I:259; I:303, 6-vol. ed.; I:279, 3rd ed.

12. Ibid. I:260; I:303, 6-vol. ed.; I:280, 3rd ed.

13. Ibid. I:260; I 304, 6-vol. ed.; I:280, 3rd ed.

> Those of the fourth Round have added *earth* as a state of matter to their stock as well as the three other elements in their present transformation. In short, none of the so-called elements were, in the three preceding Rounds, as they are now.[14]

The consideration of śloka 6 is now resumed. *The Secret Doctrine* directs attention to the fact that supervision is needed in connection with the production of suitable forms for the human kingdom:

> Thus physical nature, when left to herself in the creation of animal and man, is shown to have failed. She can produce the first two and the lower animal kingdoms, but when it cômes to the turn of man, spiritual, independent and intelligent powers are required for his creation, besides the "coats of skin" and the "Breath of animal Life." The human Monads of preceding Rounds need something higher than purely physical materials to build their personalities with, under the penalty of remaining even below any "Frankenstein" animal.[15]

The next portion of śloka 6 indicates that Superior Beings were in readiness. But because they had not supervised the processes which had taken place on Earth, it was clearly demonstrated that "Nature unaided fails." Because of the lack of supervision, the water-men were produced. So the continuation of śloka 6, together with ślokas 7 and 8, relates the consequent action which was taken by the Regents; and this is told in Chapters 10 and 11.

---

14. Ibid. I:252-3, I:296-7, 6-vol. ed.; I:272-3, 3rd ed.

15. Ibid. II:56; III:66-7, 6-vol. ed.; II:59, 3rd ed.

10.

THE BROTHERS OF THE FIFTH

THE FIRST portion of śloka 6 of the second Stanza of Dzyan was considered in the previous chapter. The second portion of śloka 6 belongs with śloka 7, which deals with the subject of the present chapter. These two ślokas describe the activities of the superior beings, called in the Stanzas the Dhyānis (which is an abbreviation of the term Dhyāni-Chohans). They are also called Watchers, because of the supervision which they maintain over the lower kingdoms, even though no humanity was then functioning on the Earth. Therefore no human evolutionary development was then in progress on the fourth globe of the Earth planetary system (Globe D), because it was in the state of dormancy or obscuration described in previous chapters.

The second portion of śloka 6 tells of the descent of the Dhyānis:

> The Dhyāni came and looked. . . . The Dhyāni from the bright Father-Mother, from the white (*Solar-lunar*) regions they came, from the abodes of the Immortal-Mortals.[1]

This would seem to imply that these Superior Beings were not stationed on the Earth but on superior globes of the Earth system. A note is added telling that the Dhyānis are Gods and planetary Spirits, especially the Ribhus. "The three Ribhus who yet become 'thrice seven in number' of their gifts."[2]

---

1. SD II:55; III:65, 6-vol. ed.; II:58, 3rd ed.

2. Ibid.

The Ribhus are a very interesting class of superior beings, even as described in exoteric literature. Their name is derived from a verbal root *rabh,* meaning clever, skillful, inventive. The Ribhus are regarded as three semidivine beings—Ribhu, Vāja, and Vibhvan—with the name of the first applied to all three. They are described as dwelling in the solar sphere. In the mythological account they are said to have formed the horses of Indra and the chariot for the Aśvins, the two divinities who appear in the sky at dawn. Indra is a Vedic deity, the king of the sidereal deities, the god of the firmament. This mythological description fits in with the "white regions" from which the Dhyānis came, especially as H. P. Blavatsky added the words "Solar-lunar" after "the white" regions. Moreover, because three Ribhus are mentioned, the suggestion is offered that they came from the three superior globes of the Earth Chain—Globes A, B, and C. "The thrice seven in number" may be explained in this way: As there are seven Watchers for the "seven groups of monads"[3]—one Watcher for each one of the seven groups—the "thrice seven" again refer to the three groups of seven on the three superior globes.

The last phrase of śloka 6: "from the abodes of the Immortal-Mortals" is significant. Again the abodes refer to the superior globes of the Earth system, inasmuch as the Immortal-Mortals represent the Śishtas, beings described as "divine mortals" and as the "future seed of humanity."[4] The śloka thus reiterates that the Dhyānis came from superior globes because the Śishtas pertaining to the human kingdom are indeed stationed on superior globes of the Earth planetary system.

What the Watchers saw when they descended upon the Earth is narrated in śloka 7. This is one of the most important ślokas, for it provides a very significant clue. It would seem that this clue has been overlooked, possibly because it was not mentioned in the comments made on

---

3. "The seven groups of monads" are explained in connection with Stanza IV, śloka 15, in Chapter 14.

4. SD II:6; III:20, 6-vol. ed.; II:6, 3rd ed. The role of the Śishtas was explained in Chapter 1.

śloka 7 in *The Secret Doctrine,* even though a noteworthy commentary from the Book of Dzyan was included there. It should be borne in mind that the second series of Stanzas from the Book of Dzyan deals solely with the Fourth Round phase pertaining to Globe D of the Earth Chain and that the first fifteen ślokas of this series cover a period prior to the appearance of the human kingdom on our Earth. In fact the "human monads" are described as being in the "spheres of expectancy"—in what may be regarded as a quasi-nirvānic state, strictly nonevolutionary.

No Fit Rūpas for Our Brothers of the Fifth

When the Dhyānis had made their descent upon this earth, they were astonished to see the creatures which had been brought forth by the efforts of the Earth without the assistance of higher beings. It is useful to include here the following explanation:

> Thus physical nature, when left to herself in the creation of animal and man, is shown to have failed. She can produce the first two and the lower animal kingdoms, but when it comes to the turn of man, spiritual, independent and intelligent powers are required for his creation, besides the "coats of skin" and the "Breath of animal Life." The human Monads of preceding Rounds need something higher than purely physical materials to build their personalities with, under the penalty of remaining even below any "Frankenstein" animal.[5]

The "first two" kingdoms represent the mineral and plant kingdoms; "the lower animal" refers to the non-mammalian animals; the mammals succeeded the first humans on our present Earth. The "coats of skin" refer to the physical vesture, which represented the third phase of human evolution on this sphere. The "breath of animal life" means Prāna—the third of the seven principles composing the human constitution. The reference to the human monads of preceding Rounds signifies the humans who had been accomplishing the first three Rounds of the Circle of Neces-

5. SD II:56; III:66-7, 6-vol. ed.; II:59, 3rd ed.

sity on this Earth. The "Frankenstein animal" refers to Mary Shelley's *Frankenstein,* the story of a medical student who fashions a man-like monster who slays his maker.

Here are the words of the Dhyānis as related in śloka 7:

> Displeased they were. Our flesh is not there. No fit rūpas for our brothers of the fifth. No dwellings for the lives. Pure waters, not turbid, they must drink. Let us dry them.[6]

The celestial Dhyānis who had descended from the superior spheres were not at all satisfied with the rūpas (forms) which they beheld. The reason was given: "Our flesh is not there." But "our flesh" does not signify physical matter inasmuch as the celestial beings would not have physical rūpas. Further clarification is given in *The Secret Doctrine* by the words "spiritual plasm":

> . . . or the fluid that contains the five lower principles of the six-principled Dhyān—and you have the secret, if you are spiritual enough to understand it.[7]

The Dhyān here signifies a representative of the Dhyāni-Chohanic kingdom who functions in a supervisory capacity over the seven kingdoms below its own hierarchy. The Dhyān is termed "six-principled," since it is represented as not having a physical vesture—Sthūla-śarīra—included in the enumeration of the seven principles. The five lower principles here signify Buddhi, Manas, Kāma, Prāna, Linga-śarīra. These five constituent principles, along with man's seventh, Ātman, are linked together by means of the spiritual plasm. This plasm, which is also termed the Dhyāni-Chohanic fluid, provides a prototypal pattern which is requisite for the development of human qualities and characteristics during the present Fourth Round stage of evolutionary development. This could not be provided by Nature's unaided efforts. Furthermore, the physical plasm is supplied by means of the physical body. Consequently the Dhyānis declared: "This is no fit rūpa for our brothers

6. Ibid. II:16; III:28, 6-vol. ed.; II:17, 3rd ed.

7. Ibid. I:224; I:270, 6-vol. ed.; I:244, 3rd ed.

of the fifth." That the rūpas are not fit has been sufficiently stressed. It is the next phrase, "our brothers of the fifth," that deserves special notice.

The Significance of "Our Brothers of the Fifth"

The very first word is important. It expresses the idea that there is an intimate kinship between the Dhyānis and the beings for whom the rūpas are intended—so much so that they are regarded as "brothers." There is a kinship because of the close relationship between the two kingdoms, as well as the fact that both classes of beings originated from the same primordial Source.

The relationship is mentioned in a significant passage:

> To the highest, we are taught, belong the seven orders of the purely divine Spirits; to the six lower ones belong hierarchies that can occasionally be seen and heard by men, and who do communicate with their progeny of the Earth; which progeny is indissolubly linked with them, each principle in man having its direct source in the nature of those great Beings, who furnish us with the respective invisible elements in us.[8]

However, the question arises as to why the words "our brothers of the fifth" are used, instead of "brothers of the fourth"—because it is the Fourth Round that is in process of evolutionary development. Moreover, it is the peopling of the *fourth* globe of the Earth planetary system that is being discussed. Here are some points to be considered:

1. The "brothers of the fifth" cannot mean the monads who will be undertaking the Fifth Round, because the Stanzas have introduced us only to Rounds, One, Two, and Three. Furthermore our present human evolutionary life-cycle is now only in its *Fourth Round.*

> Its [our Earth's] Humanity develops fully only in the Fourth—our present Round.[9]

8. Ibid. I:133; I:191, 6-vol. ed.; I:158, 3rd ed.

9. Ibid. I:159; I:213, 6-vol. ed.; I:183, 3rd ed.

2. The words cannot be applied to the fifth globe of the Earth planetary chain, because the very first words of the first Stanza mentions "the Lha which turns the fourth," which signifies the fourth globe of the system.

3. The "brothers of the fifth" cannot be used to describe the fifth Root-Race, because the Stanza has just been referring to the water-men "terrible and bad" representing a stage prior to the birth of the *First* Root-Race—a time-period of some millions of years ago.

4. Could the "brothers of the fifth" relate to the fifth kingdom (regarded as the superhuman kingdom)? Not so, because the human kingdom is the *fourth* of the four manifested kingdoms: the mineral, the plant, the animal and the human. Of the kingdoms described in *The Secret Doctrine* as comprising the hierarchical Ladder of Life, the human kingdom is really the seventh, not the fourth; because there are three kingdoms mentioned below the mineral kingdom called the three elemental kingdoms. Furthermore, there are three kingdoms superior to the human kingdom, termed the Dhyāni-Chohanic kingdoms.

5. Could the "brothers of the fifth" apply to the fifth principle of man's sevenfold constitution—to Manas, his mind-principle? This is plausible but it can readily be shown that it is unacceptable, because it requires one Round for the full development of *one* of the principles of man's sevenfold constitution, and there have been only three Rounds during which the three lowest principles—Prāna, Linga-śarīra, Śthūla-śarīra, or the vital principle, the etheric double, and the physical body—have been developed.

> It is not in the course of natural law that man should become a perfect septenary being, before the seventh race in the seventh Round. Yet he has all these principles latent in him from his birth. Nor is it part of the evolutionary law that the Fifth principle (*Manas*), should receive its complete development before the *Fifth* Round.[10]

---

10. Ibid. II:167; III:176, 6-vol. ed.; II:177, 3rd ed.

Thus the complete evolutionary development of the fourth principle, Kāma, will be completed at the conclusion of the present circuit—the Fourth Round. The complete evolutionary development of Manas, the mind-principle, will be concluded at the end of the Fifth Round.

It should be borne in mind that śloka 7 relates what occurred *before* the beginning of the peopling of the Earth (or Globe D), at a time prior to the First Root-Race. It is not describing what rūpas (forms) are required for the development of the fifth principle, for the evolution of that principle will take place during the Fifth Round of human evolutionary development. Nor is the fifth globe (Globe E) of the Earth Chain applicable. The rūpas under consideration in the seventh śloka are those which should be appropriate to, and applicable to, Fourth Round conditions. It is the midway Round, and the śloka describes a preparatory state or condition of the fourth globe of the Earth planetary system.

Then with regard to the cosmic Element-Principles, or Tattvas, which undergo evolutionary development during each Round, it is the fourth Element-Principle, called the element of Earth or the Prithivī-tattva which is being developed during the present Fourth Round.

Only one interpretation of "the brothers of the fifth" seems to fit in with the passages quoted from the Commentaries on the Book of Dzyan, but *The Secret Doctrine's* definition of the planes should first be considered.

The Significance of Planes

A word of caution must be given regarding the usual diagrammatic presentation of the seven planes, showing the planes as layers situated one above the other. This has led to some confusion. H. P. Blavatsky was emphatic on this point:

> . . . please do not allow your fancy to suggest to you layers like strata or beds laid one over the other, for this would only lead to another absurd misconception.[11]

11. *The Key to Theosophy,* (hereafter cited as *Key*) p. 88 (original edition).

A plane was thus defined:

> From the Latin *planus* (level, flat), an extension of space, whether in the physical or metaphysical sense. In Occultism, the range or extent of some state of consciousness, or the state of matter corresponding to the perceptive powers of a particular set of senses or the action of a particular force.[12]

With reference to the six invisible companion globes of a planetary system situated on the planes superior to that of our physical Earth, which is located on the lowest cosmic plane, it is stated that these six globes are:

> . . . outside our physical means of perception, or plane of being. It is not only that their material density, weight, or fabric are entirely different from those of our earth and the other known planets; but they are (to us) on an entirely different *layer* of space, so to speak; a layer not to be perceived or felt by our physical senses. . . . What I mean by "layer" is that plane of infinite space which by its nature cannot fall under our ordinary waking perceptions, whether mental or physical; but which exists in nature outside of our normal mentality or consciousness, outside of our three dimensional space, and outside of our division of time. Each of the seven fundamental planes (or layers) in space—of course as a whole, as the pure space of Locke's definition, not as our finite space—has its own objectivity and subjectivity, its own space and time, its own consciousness and set of senses.[13]

The location of the superior six globes of the earth planetary system are thus defined: Globe A and Globe G are situated on the fourth cosmic plane; Globe B and Globe F on the fifth; Globe C and Globe E on the sixth. The lowest plane, the seventh, on which our Earth (Globe D) alone is located, is the physical or terrestrial plane.

Another factor of great significance in connection with the planes is that *each of the seven cosmic planes is itself*

---

12. *Key* p. 360.

13. Ibid. pp. 88-9.

divisible into "seven planes." These seven subdivisions are often referred to as "subplanes."

Concerning these subdivisions of the planes, and using the lowest plane (enumerated as the seventh cosmic plane), as an example, H. P. Blavatsky stated:

> Taking the lowest only, the Terrestrial (it was afterwards decided to call this plane Prākritic), it is divisible into seven planes, and these again into seven, making the forty-nine.[14]

In this enumeration the first subdivision is regarded as the loftiest, and the seventh as the lowest. The subdivisions of the seventh Cosmic Plane, termed the Terrestrial, are enumerated in this manner:[15]

	Subplane	
Terrestrial or Seventh Cosmic Plane	1.	Para-Ego or Ātmic
	2.	Inner Ego or Buddhic
	3.	Ego-Manas
	4.	Kāma-Manas or Lower Manas
	5.	Prānic-Kāma or Psychic
	6.	Astral
	7.	Objective

The Evolutionary Development of Planetary Systems

Since the subdivisions of the cosmic planes have been explained, the evolutionary development of planetary systems may now be presented, and it is this which provides the key to the significance of "the brothers of the fifth."

The Secret Doctrine makes two postulates concerning planetary systems:

> 1. Everything in the metaphysical as in the physical Universe is septenary. Hence every sidereal body, every planet, whether visible or invisible, is credited with six companion globes. The evolution of life proceeds on these seven globes or bodies from the 1st to the 7th in Seven Rounds or Seven Cycles.

---

14. SD V:524, 6-vol. ed.; III:551, 3rd ed.

15. Cf. SD V:525, 6-vol. ed.; III:552, 3rd ed.

> 2. These globes are formed by a process which the Occultists call the "rebirth of planetary chains (or rings)." When the seventh and last Round of one of such rings has been entered upon, the highest or first globe "A," followed by all the others down to the last, instead of entering upon a certain time of rest—or "obscuration," as in their previous Rounds—begins to die out. The "planetary" dissolution (pralaya) is at hand, and its hour has struck; each globe has to transfer its life and energy to another planet [globe].[16]

As explained in earlier chapters, the seven globes which form a planetary system are designated by the first seven letters of the alphabet: A, B, C, D, E, F, G. The phrase, "the highest or first globe 'A,' followed by all the others down to the last," signifies that the remaining six globes transfer their energies one by one sequentially each to a laya-center,[17] until the last one is accomplished. Thus Globe G is the last of the manifested globes to enter into the state or condition of pralaya. In other words the entrance into pralaya is not made by means of the situation of the globes on their respective planes,[18] but by means of evolutionary emergence and developmental phases, the latter being governed by means of the development of the ten kingdoms, or life-waves, which evolve on each of the seven globes.

Diagram No. 7 is intended to represent the seven manifested globes of the Moon Chain and the transference of its life and energies to the Earth planetary system's seven globes. The seven globes of the lunar planetary system are designated alphabetically A to G, and each globe is linked to the Earth Chain's globes designated A+ to G+. Globe D represents the Moon which is the satellite of the Earth (Globe D+). When the life-period of the lunar system had run its course and the state of pralaya was entered, each globe

---

16. SD I:158-9; I:213, 6-vol. ed.; I:182, 3rd ed.

17. A Laya-Center or Laya-Point signifies that state where substance becomes homogeneous and is unable to act or differentiate.

18. In the previous section entitled "The Significance of the Planes," the assignment of the seven globes of a planetary system to specific cosmic planes was described.

The Moon Chain

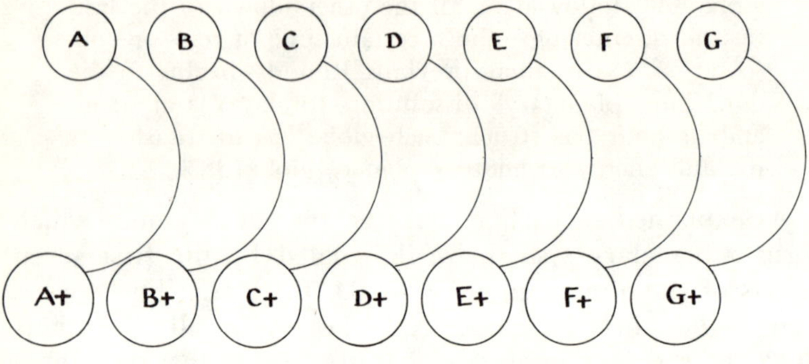

The Earth Chain

DIAGRAM 7

of the Moon Chain transferred its inner (or higher) principles into a Laya-Center. In due time these inner principles awakened into activity and became manifest as the Earth Chain.

The globes of the Moon Chain which were formerly on the fourth cosmic plane—termed Globes A and G—transferred their life and energy to a Laya-Center in the same cosmic plane; Globes B and F to the fifth cosmic plane; Globes C and E to the sixth cosmic plane and Globe D to the seventh cosmic plane. When the cyclical period of the Pralaya was completed and the time for the awakening of the new planetary system had arrived, Globe A was the first of the manifested globes to spring into activity. This awakening was followed in serial order by Globes B, C, D, E, F, and finally G. That is to say the rebirth of a planetary system does not occur by means of the sequence of planes (i.e., *not* by the awakening of Globes A and G, followed by B and F, C and E, and finally D) but by means of the Descending and Ascending Arcs of evolution. In other words it is done by the process of Pravritti and Nivritti. Pravritti signifies the process of unfolding of inherent potencies resulting in the evolution of forms, whereas Nivritti means the process of infolding of forms and an evolution of spirit.

Attention is directed to this factor. In the diagrammatic representation of the Cosmic Planes and the situation of the seven globes as portrayed on the Fourth, Fifth, Sixth, and Seventh Cosmic Planes, the seven globes are represented from the standpoint of *emanation*. This signifies that the First Cosmic Plane is regarded as the *highest*, portraying the point of emanation, whereas the Seventh Cosmic Plane represents the *lowest* and most physical aspect. Consequently Globes A and G are positioned on Cosmic Plane No. 4 and the Earth is placed on Cosmic Plane No. 7—the most material phase of the evolutionary process.

However, when depicting the subplanes in connection with the Cosmic Planes, the subplanes are enumerated by means of their *evolutionary* status, consequently subplane No. 1 represents the first or lowest degree of evolutionary development, whereas subplane No. 7 represents the loftiest aspect of evolutionary development. Therefore the lunar planetary system is represented as functioning on the Fourth subplane of the Seventh Cosmic Plane, whereas the Earth planetary system is represented as performing its evolution on the Fifth subplane of the Seventh Cosmic Plane—in conformity with the statement made in *The Secret Doctrine* that the Moon Chain is inferior to the Earth Chain.

Diagram No. 8 is intended to represent the positions of the respective seven globes of the Moon Chain and the Earth Chain. On the left-hand side of the diagram the seven globes of the Moon Chain are portrayed: Globes A and G are situated on the 4th Cosmic Plane; Globes B and F on the 5th Cosmic Plane; Globes C and E on the 6th Cosmic Plane; Globe D, our Moon, alone on the 7th Cosmic Plane. On the right-hand side, the Earth Chain's seven globes are similarly situated. Each one of the Cosmic Planes is subdivided into seven subplanes. The seven globes of the Moon Chain are shown on the 4th subplanes of the respective Cosmic Planes. The seven globes of the Earth Chain are placed on the *5th subplanes* of the respective Cosmic Planes. This

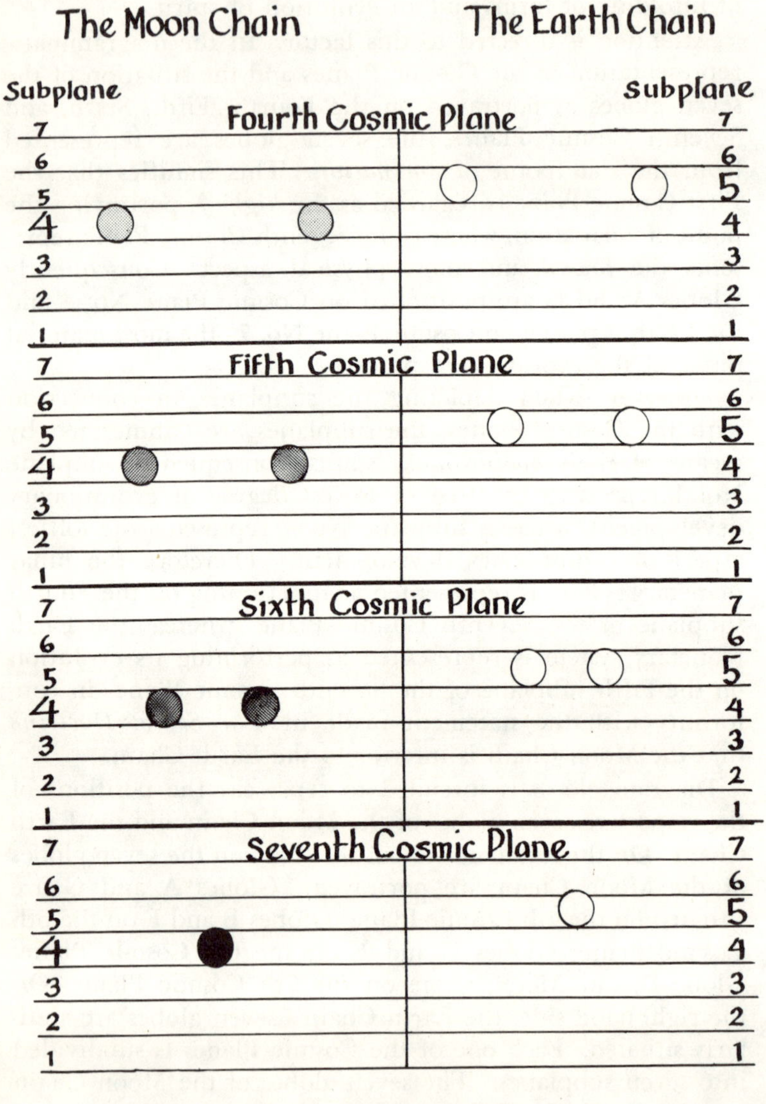
The Moon Chain
The Earth Chain
Subplane
Subplane
Fourth Cosmic Plane
Fifth Cosmic Plane
Sixth Cosmic Plane
Seventh Cosmic Plane
7
6
5
4
3
2
1

DIAGRAM 8

represents an evolutionary advancement of the Earth planetary system. Thus the Earth Chain is one subplane higher than the Moon Chain; each one of the globes of the Earth planetary system is so represented. Consequently the evolutionary development of all the ten kingdoms, or life-waves, on the Earth Chain is at a stage superior to that of the Moon Chain.

The law of periodicity is applicable to planetary and solar systems, as well as to human beings. In the planetary system a planetary embodiment is equivalent to a planetary manvantara, and this fulfills a seven-round cyclic period of activity. When a seven-round cycle of activity has been completed, one by one the seven globes of the system pass into pralaya in sequential order. Upon the completion of the pralaya, the reawakening of the planetary system enables all the globes of the system to function on a subplane superior to that of the previous embodiment. The new planetary system is brought into birth by means of a transfer of the life and energy—or the inner principles—of the old planetary chain, to a laya-center. Thus the stimulus for reawakening was inherent in this laya-center awaiting the appropriate cyclic time-period for remanifestation.

Diagram No. 9 is intended to portray the evolutionary advancement made each time a planetary system reappears after a pralaya. The Seventh Cosmic Plane alone is represented, and divided into seven subplanes. The globes on subplanes 1, 2, and 3 represent prelunar planetary systems. The symbol of the Moon is shown on the 4th subplane of the Seventh Cosmic Plane. The symbol of the Earth is shown on the 5th subplane of the Seventh Cosmic Plane—representing the Earth Chain's evolutionary advancement from the Moon Chain (in the reembodiment of the planetary system).

Since it has been stated that the Moon Chain is inferior to the Earth Chain, one may postulate that there have been four previous embodiments, or births, of the planetary system of which the Earth is now the representative, functioning on the Seventh Cosmic Plane of the fifth embodiment or rebirth.

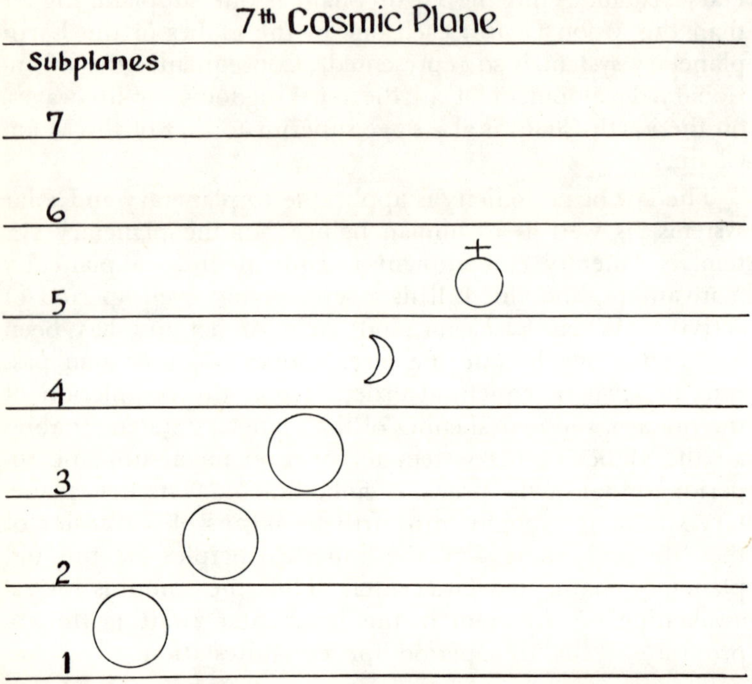

DIAGRAM 9

The statement made in the 7th śloka of the second Stanza of Dzyan is now made apparent: "No fit rūpas for our brothers of the fifth" signifies that the forms produced without the assistance of the Dhyāni-Chohanic fluid are not suitable for the monads of the human kingdom applicable to the fifth embodiment, or *fifth rebirth of the planetary system.*

All planetary systems or chains have their evolutionary cycles to accomplish. Therefore, a planetary chain's evolution follows a cyclical process similar to that of the evolution of a globe. This is clearly stated in *The Secret Doctrine* in a third postulate:

> 3. Our Earth, as the visible representative of its invisible superior fellow globes, its "lords" or "principles,"

> has to live, as have the others, through seven Rounds. During the first three, it forms and consolidates; during the fourth it settles and hardens; during the last three it gradually returns to its first ethereal form; it is spiritualized, so to say.[19]

Thus during the life-cycle of a planetary system, consisting of a seven-round cyclic period of activity, the individual globes carry on their evolutionary development by pursuing the course of a Cycle of Descent, described as the period of the formation and consolidation of the globes, and a corresponding subsequent Cycle of Ascent.

On the theme of the evolution of the superior kingdoms of the Earth—referred to as the Celestial Hierarchy or Dhyān Chohans—in connection with the reembodiments of "worlds," or globes, H. P. Blavatsky states:

> A Dhyān Chohan has to become one; he cannot be born or appear suddenly on the plane of life as a full-blown angel. The Celestial Hierarchy of the present Manvantara will find itself transferred in the next cycle of life into higher, superior worlds, and will make room for a new Hierarchy, composed of the elect ones of our mankind.[20]

By analogy, then, it was during the prelunar planetary system that the Agnishvātta Pitris (who compose one of the Dhyāni-Chohanic kingdoms) accomplished their cyclical evolution in the most superior of the *manifested* kingdoms on the hierarchical Ladder of Life, that is to say the kingdom equivalent to our human kingdom, on that prelunar Chain.

> As from the highest Archangel (Dhyān Chohan) down to the last conscious "Builder" (the inferior class of Spiritual Entities), all such are *men,* having lived aeons ago, in other Manvantaras, on this or other Spheres; so the inferior, semi-intelligent and non-intelligent Elementals—are all *future* men. That fact alone—that a Spirit is endowed with intelligence—is a proof to the

---

19. SD I:159; I:213, 6-vol. ed.; I:183, 3rd ed.

20. Ibid. I:221; I:268, 6-vol. ed.; I:242, 3rd ed.

> Occultist that that Being must have been a *man,* and acquired his knowledge and intelligence throughout the human cycle.[21]

Here is another pronouncement which indicates that there exists a relationship between men and the Agnishvātta Pitris, also called Fire Dhyānis, meaning the Dhyāni-Chohans who awakened the "fire of Mind":

> That class of the "Fire Dhyānis," which we identify on undeniable grounds with the Agnishvāttas, is called in our school the "Heart" of the Dhyān-Chohanic Body; and it is said to have incarnated in the third race of men and made them perfect. The esoteric Mystagogy speaks of the mysterious relation existing between the hebdomadic essence or substance of this angelic Heart and that of man, whose every physical organ, and psychic, and spiritual function, is a reflection, so to say, a copy on the terrestrial plane of the model or prototype *above.*[22]

Notice the words: "the mysterious relation existing between the Agnishvātta Pitris and men." Since the Agnishvāttas are the Heart of the Dhyān-Chohanic Body, these Superior Beings ascended to their appropriate "station" on the hierarchical Ladder of Life after they had performed their work of enlightenment.[23]

Following the conclusion of the pralaya of the prelunar system or the Third Planetary Chain, the next rebirth of the planetary system brought about the embodiment of the Fourth or Lunar Planetary Chain, of which the present Moon was the fourth globe. During the seven-round cycle of existence which signifies the Lunar Chain Manvantara, the Lunar Pitris achieved their evolutionary development

---

21. Ibid. I:277; I:320, 6-vol. ed.; I:297, 3rd ed.

22. SD II:91-2; III:100, 6-vol. ed.; II:96, 3rd ed.

23. Here is another statement showing that the Mind-born Sons (i.e. Mānasaputras or Agnishavātta-Pitris) passed through the evolutionary stage of the human kingdom in a preceding planetary system: "The Mind-born Sons, the Rishis, the Builders, etc., were all men—of whatever forms and shapes—in other worlds and the preceding Manvantara." (SD I:107; I:167, 6-vol. ed.; I:132, 3rd ed.)

of Seven Rounds. This accomplished their graduation on the Lunar Chain from what was there the equivalent of the present human kingdom.

When the seven-round period of activity of the Moon Chain was concluded the inner principles of each globe of the system were transferred, one by one, to a laya-center. Then the pralaya of the Lunar Chain ensued.

When the Night of Brahmā was concluded (which signifies the close of a pralaya), the Fifth Planetary Chain, which is that of the present Earth Chain, came to birth. It represents the rebirth of the Lunar Chain. All the seven globes of the Earth Chain are situated one subplane *above* the subplane on which the Lunar Chain functioned. Therefore the Moon Chain is *inferior* to the Earth Chain. Although the Lunar Pitris are represented as having successfully achieved their seven-round cycle of evolutionary development on the Lunar Chain, nevertheless they are *not yet* sufficiently "liberated" from the lunar influence; consequently they are not yet actively functioning in the evolution of the Dhyāni-Chohanic Kingdom.

The Lunar Chain Is Inferior to the Terrestrial Chain

The Secret Doctrine emphatically states that the Moon Chain is inferior to the Earth Chain. This is clearly shown by five quotations now to be given. The first is a significant passage from the Volume I. Attention is drawn to the "occult maxim"; also to the closing sentence, which answers our original question: "Where do people come from?"

> A few words more of the Moon:
>
> [1] What, it may be asked, are the "Lunar Monads"? . . . It must be plain to everyone that they are Monads, who, having ended their life-cycle on the lunar chain, which is inferior to the terrestrial chain, have incarnated on this one. But there are some further details which may be added, . . . it may be stated that our satellite is only the gross body of its invisible principles. Seeing then that there are 7 Earths, so there are 7 Moons, the last one alone being visible; the same for the Sun, whose

> visible body is called a Māyā, a reflection, just as man's body is. "The real Sun and the real Moon are as invisible as the real man," says an occult maxim.
>
> And it may be remarked *en passant* that those ancients were not so foolish after all who first started the idea of "the seven moons" . . . .
>
> It is, then, the Moon that plays the largest and most important part, as well in the formation of the Earth itself, as in the peopling thereof with human beings.[24]

The occult maxim which states that the real sun, the real moon, and the real man are invisible to ordinary sight, signifies that these are "inner principles." Thus the real man is the immortal portion or component which consists of the Monad (Ātma-Buddhi) in conjunction with the Reincarnating Ego (Higher Manas).

The Lunar Pitris are described in *The Secret Doctrine* as planetary spirits because of their having achieved the full seven rounds of evolutionary development in the Lunar Chain:

> [2] These are neither gods, nor supernatural Beings, but advanced Spirits from another and lower planet, reborn on this one, and giving birth in their turn in the present Round to present Humanity.[25]

The theme of the Lunar Pitris and the vehicles which they produced on earth is stressed in this quotation:

> [3] The Moon being an inferior body to the Earth even, to say nothing of other planets, the terrestrial men produced by her sons—the lunar men or "ancestors"—from her shell or body, cannot be immortal.[26]

In the light of the foregoing the Earth Chain cannot properly be regarded as the fourth in the series of chain embodiments with the Moon Chain as the third in the series, as has sometimes been done, because that would make the

---

24. SD I:179-80; I:231-2, 6-vol. ed.; I:202-3, 3rd ed.

25. Ibid. II:611; IV:182, 6-vol. ed.; II:646, 3rd ed.

26. Ibid. II:45; III:56, 6-vol. ed.; II:48, 3rd ed.

Terrestrial Chain inferior by being on a lower cosmic subplane to the Lunar Chain, instead of being *superior* as previously stated.

> [4] Every such chain of worlds is the progeny and creation of another, *lower,* and *dead* chain—*its reincarnation,* so to say.[27]

In a passage in the first volume of *The Secret Doctrine,* which deals with the coming into being of worlds and systems of worlds, a corroborative reference is made to the fact that each time there is a rebirth of a planetary system it is on a higher subplane. Incidentally, the same quotation refers also to "the Luminous Sons of Manvantaric Dawn," who are, indeed, the "Watchers" or Regents of the seven globes of a planetary system. The "Age of Brahmā" signifies the life-cycle of a solar system:

> [5] . . . in this doctrine, which deals only with our Planetary System, they [the Sons of Manvantaric Dawn], as the architects of the latter, are also called the "Watchers" of the Seven Spheres, which exoterically are the Seven planets, and esoterically the seven earths or spheres (planets) [i.e. globes] of our chain also. The opening sentence of Stanza I of the first volume, when mentioning "Seven Eternities," is made to apply both to the *Mahā-Kalpa,* or "the (great) Age of Brahmā," as well as to the Solar pralaya and subsequent resurrection of our Planetary System on a higher plane.[28]

The Dwellings for the Monads

The narrative presented in śloka 7 of Stanza II is now resumed. The complaint that the rūpas prepared by unaided Nature were not fit for the "brothers of the fifth" is further emphasized by the next statement of the Dhyānis: "No dwellings for the lives." The dwellings are, of course, the unsatisfactory rūpas. The monads are "the lives." The Sanskrit word for the "lives" is *jīvas,* often used in place of

27. Ibid. I:152; I:207, 6-vol. ed.; I:176, 3rd ed.

28. Ibid. I:53; I:124, 6-vol. ed.; I:83-4, 3rd ed.

monads in the commentaries. An explanatory footnote is added after the word "lives": "The Monads of the *presentments* of men of the *Third* Round."[29] The monads (lives or jīvas) here referred to have accomplished three complete circuits of evolutionary development on each of the seven globes, which is equivalent to the accomplishment of three complete Rounds. They have also completed seven major stages of evolutionary development on Globe C, corresponding to our Root Race periods. These monads are awaiting entry onto Globe D, hence they are termed "presentments of men of the Third Round."

The next two sentences of the śloka are in symbolic language: "Pure waters, not turbid, they must drink. Let us dry them (the waters)." Here "the waters" suggest that reference is made to third round conditions; likewise that the rūpas which had been brought into manifestation without the supervision of the superior "Lords," as well as the substance of the prevailing "waters" were not satisfactory for life generally on the new globe, nor for the jīvas (monads) who were to inhabit it. It should be borne in mind that during the Third Round the Element-Principle of Water (Apas-tattva) was predominant. Consequently, a rūpa in which water was preponderant would be in keeping with Third Round conditions. However, this would not apply to the Fourth Round, during which the Element-Principle of Earth (Prithivī-tattva) predominates.

This explains why the rūpas which had been brought forth were described in the sixth śloka as "the water-men terrible and bad." It was because the forms represented rūpas pertaining to the cosmic element-principle of Apas-tattva instead of those which would be suitable to the Prithivī-tattva. Consequently these were not "fit rūpas for our brothers of the fifth."

---

29. Ibid. II:57; III:67, 6-vol. ed.; II:60, 3rd ed.

11.

THE ACTIVITIES OF SUPERIOR BEINGS

IN THE account of the peopling of the Earth, the archaic manuscript states that none of the kingdoms of nature are left without their Watchers or Guardians. These supervisors provide assistance, especially when most needed.

The Earth—the fourth sphere in the series of seven globes —represents the critical point in the cyclical evolution of this planetary system. This is so because the Fourth Round —which is the present period of the Earth's life-cycle—is the lowest point in the life-cycle of the planetary system, from the standpoint of its evolution. It is termed the lowest point in the Arc of Descent. The Fourth Round is the midway Round. Three Rounds of evolution have been completed and, when the Fourth Round is completed, three more Rounds must be accomplished in order to fulfill the evolutionary development of the Earth planetary system, under the guidance of the Regents of the Earth.

The Coming of the Dhyānis

In the narrative of the Stanzas of Dzyan the point has been reached where the Dhyānis, the Regents of the Earth planetary system, have been called from "the white regions" to view the results of Nature's unaided efforts to produce rūpas for the monads waiting to people the Earth. Upon seeing "the water-men, terrible and bad," they declared that such forms were "no fit rūpas for our brothers of the fifth. Let us dry the waters." The next series of slokas tells of the activities of the Regents. Śloka 8 opens with these descriptive words:

> The flames came. The fires with the sparks; the night fires and the day fires. They dried out the turbid dark waters. With their heat they quenched them. The Lhas (Spirits) of the High, the Lhamayin (those) of below, came. They slew the forms which were two- and four-faced. They fought the goat-men, and the dog-headed men, and the men with fishes' bodies.[1]

Regarding the three first sentences *The Secret Doctrine* comments:

> The "Flames" are a Hierarchy of Spirits parallel to, if not identical with, the "burning" fiery *Saraph* (Seraphim) mentioned by Isaiah (vi: 2-6), those who attend, according to Hebrew Theogony, "the Throne of the Almighty." Melha is the Lord of the "Flames." When he appears on Earth, he assumes the personality of a Buddha, says a popular legend. He is one of the most ancient and revered *Lhas,* a Buddhist St. Michael.[2]

Here is Isaiah's description of the Seraphim, standing above the Lord's throne:

> . . . each one had six wings; with twain he covered his face, and with twain he covered his feet, and with twain he did fly. (vi:2)

Melha is the Tibetan god of fire; the Lhas are the Tibetan planetary spirits. As described in the śloka, the Lhas of the high signify the superior planetary beings. On the other hand the Lhamayin signify those who are below, for the literal meaning of the Tibetan word means "below the Lhas." Exoterically the Lhamayin are one of the six classes of celestial beings represented as living on the slopes of Mount Sumeru. Thus the Lhamayin are lesser planetaries, as H. P. Blavatsky points out:

> The word "Below" must not be taken to mean infernal regions, but simply a spiritual, or rather ethereal, Being

---

1. SD, II:63; III:74, 6-vol. ed.; II:67, 3rd ed.

2. Ibid. II:63; III:74, 6-vol. ed.; II:67, 3rd ed.

> of a lower grade, because nearer to the Earth, or one step higher than our terrestrial sphere; while the Lhas are Spirits of the highest Spheres.[3]

Then, in further reference to śloka 8, an explanation is provided in regard to the "Flames":

> Besides a statement of a purely physical nature and belonging to the evolution of life on Earth, there may be another allegorical meaning attached to this śloka, or indeed, as is taught, several. The FLAMES, or "Fires," represent Spirit, or the male element, and "Water," matter, or the opposite element. And here again we find, in the action of the Spirit, slaying the purely material form, a reference to the eternal struggle, on the physical and psychic planes, between Spirit and Matter, besides a scientific cosmic fact.[4]

In connection with the last two sentences of śloka 8, referring to the monsters which had been produced in the waters, *The Secret Doctrine* poses some pertinent questions and proceeds to supply fitting responses:

> . . . what will science say to this idea of a primordial creation of species? . . . if it is nature and the physical law of evolution that are the creators of all there is now on Earth, why could there be "no such abyss" when the globe was covered with waters, in which numbers of monstrous beings were generated? Is it the "human beings" and animals with human heads and double faces, which are a point of the objection? But if man is only a higher animal and has evolved from the brute species by an infinite series of transformations, why could not the "missing links" have had human heads attached to the bodies of animals, or, being two-headed, have heads of beasts and *vice versa,* in Nature's early efforts? Are we not shown during the geological periods, in the ages of the reptiles and the mammalia, lizards with birds' wings, and serpents' heads on animal bodies. And, arguing from the standpoint of science, does not even our

3. Ibid.

4. Ibid. II:63-4; III:74, 6-vol. ed.; II:67, 3rd ed.

modern human race occasionally furnish us with monster-specimens: two-headed children, animal bodies with human heads, dog-headed babies, etc., etc.? And this proves that, if nature will still play such freaks now that she has settled for ages into the order of her evolutionary work, monsters, like those described by Berosus, were a possibility in her opening program; which possibility may even have existed once upon a time as a law, before she sorted out her species and began regular work upon them; which indeed now admits of definite proof by the bare fact of "*Reversion,*" as science puts it.

This is what the doctrine teaches and demonstrates by numerous proofs.[5]

The Moon Is Far Older Than the Earth

The ninth śloka continues the subject of the waters, expressed in symbolic language:

> Mother-water, the great sea wept. She arose, she disappeared in the Moon, which had lifted her, which had given her birth.[6]

Although one explanation is given in *The Secret Doctrine* in regard to the water composing so much of the Earth's surface, there must be some other meaning to the "great sea" which arose and disappeared in the moon. Another interpretation is offered: as the Water Element-Principle lessened, the Earth Element-Principle became more and more predominant, because the Prithivī-Tattva is the dominant Element-Principle during the Fourth Round, especially on Globe D (our Earth) ; whereas the Water Element-Principle was predominant during the Third Round.

There is also a reference to the two kinds of water mentioned in śloka 7: pure waters and turbid waters. The turbid waters would apply to the waters in the former cycle of evolution—the Third Round conditions; the pure waters relate to the present cycle. The Dhyānis assisted in dispers-

---

5. Ibid. II:55-6; III:66, 6-vol. ed.; II:58-9, 3rd ed.

6. Ibid. II:64; III:74, 6-vol. ed.; II:67, 3rd ed.

ing the turbid waters when they declared: "Let us dry them" —signifying the turbid waters. Concerning the relationship of the Earth to the Moon:

> The Moon is far older than the Earth; it is the latter which owes its being to the former, however astronomy and geology may explain the fact. Hence, the tides and the attraction to the Moon, as shown by the liquid portion of the Globe ever striving to raise itself towards its parent. This is the meaning of the sentence that "the Mother-Water arose and disappeared in the Moon, which had lifted her, which had given her birth."[7]

With regard to the tides: scientific observations have determined that at most seaside places the water attains its highest level about twice a day, at approximately 12-hour and 25-minute intervals, although this varies during a week. Then, too, the highest water is about the same distance above the mean level as the low water is below the mean level. Further observations show that the high water levels bear a relation to the positions of the moon and the sun. The 12-hour, 25-minute period is half that of the moon's apparent revolution around the earth. The length of time between the moon's crossing of the meridian of a place and the next high water at that place is known as the lunitidal interval, or the high water interval for the place. Similarly the length of time between the moon's crossing of the meridian and the next low water is termed the low-water interval. The daily inequality of the tides is always associated with the angular difference of the moon or sun north or south from the celestial equator.

Scientific observations testify to the fact that "earth tides" are created by the moon and sun. Hawaii is described as rising and falling about four inches daily. By means of the gravimeter, studies of such earth movements assist in determining the rigidity of the earth.

Furthermore, it is asserted that the moon attracts every particle of the earth and the ocean.

---

7. Ibid. II:64; III:75, 6-vol. ed.; II:68, 3rd ed.

Another significant passage relates the influence of the Moon upon the Earth:

> The Moon is now the cold residual quantity, the shadow dragged after the new body [the Earth], into which her living powers and "principles" are transfused. She is now doomed for long ages to be ever pursuing the Earth, to be attracted by and to attract her progeny. Constantly *vampirized* by her child, she revenges herself on it by soaking it through and through with the nefarious, invisible and poisoned influence which emanates from the occult side of her nature. For she is a *dead*, yet a *living body*. The particles of her decaying corpse are full of active and destructive life, although the body which they had formed in soulless and lifeless. Therefore its emanations are at the same time beneficent and maleficent.[8]
>
> In reality the Moon is only the satellite of the Earth in one respect, viz., that physically the Moon revolves round the Earth. But in every other respect it is the Earth which is the satellite of the Moon, and not *vice versa*. Startling as the statement may seem it is not without confirmation from scientific knowledge. It is evidenced by the tides, by the cyclic changes in many forms of disease which coincide with the lunar phases; it can be traced in the growth of plants, and is very marked in the phenomena of human gestation and conception. The importance of the Moon and its influence on the Earth were recognized in every ancient religion.[9]

The Earth Is Made Bare

The next śloka in the Stanzas of Dzyan, No. 10, tells what happened to the Earth when the Dhyānis had "dried out the turbid dark waters." It is followed by an explanatory passage:

> When they (the Rūpas) were destroyed, Mother-earth remained bare, she asked to be dried[10]

---

8. Ibid. I:156; I:211, 6-vol. ed.; I:180, 3rd ed.

9. Ibid. I:180; I:232, 6-vol. ed.; I:202, 3rd ed.

10. Ibid. II:65; III:75, 6-vol. ed.; II:68, 3rd ed.

The time for its incrustation had arrived. The waters had separated and the process was started. It was the beginning of a new life. This is what one key divulges to us. Another key teaches the origin of Water, its admixture with Fire (liquid fire it calls it), and enters upon an alchemical description of the progeny of the two—solid matter such as minerals and earths. From the "Waters of Space," the progeny of the male Spirit-Fire and the female (gaseous) Water has become the Oceanic expanse on Earth. Varuna is dragged down from the infinite Space, to reign as Neptune over the finite Seas. As always, the popular fancy is found to be based on a strictly scientific foundation.

Water is the symbol of the female element everywhere. . . . It is the universal matrix or the "Great Deep." Venus, the great Mother-Virgin, issues forth from the Sea-wave, and Cupid or Eros is her son. But Venus is the later mythological variant of *Gaia* (or *Gaea*), the Earth, which, in its higher aspect is Nature (*Prakriti*), and metaphysically Aditi, and even Mūlaprakriti, the root of Prakriti or its noumenon.

Hence Cupid or Love in his primitive sense is Eros, the Divine Will, or *Desire of manifesting itself through visible creation.* Thence Fohat, the prototype of Eros, becomes on Earth the great power "Life-electricity," or the Spirit of "Life-giving." Let us remember the Greek Theogony and enter into the spirit of its philosophy. We are taught by the Greeks that all things, gods included, owe their being to the Ocean and his wife Tethys, the latter being Gaia, the Earth or Nature. But who is *Ocean*? Ocean is the immeasurable SPACE (Spirit in Chaos), which is the Deity; and Tethys is not the Earth, but primordial matter in the process of formation. In our case it is no longer Aditi-Gaia who begets *Ouranos* or Varuna, the chief Āditya among the seven planetary gods, but Prakriti, materialized and localized.[11]

In referring to the deities of Hindu and Greek mythology, H. P. Blavatsky has presented an enlightening concept. Some of the Greek myths are well known, but others are not, be-

---

11. Ibid. II:65; III:75-6, 6-vol. ed.; II:68-9, 3rd ed.

cause they pertain to an older hierarchical system. There is a parallel between the transference of the "inner principles" from the Moon planetary system to the Earth planetary system, for the latter's formation, maintenance, and vitality, and the transfer of the deific principles of the older hierarchy of gods and goddesses to a later hierarchical system forming the familiar deities of Greek mythology. The deities first to be enumerated represent gods and goddesses of the presiding hierarchy: they will be followed by reference to the more ancient hierarchy.

Neptune, in Roman mythology (identified with the Greek Poseidon, son of Kronos and Rhea and brother of Zeus), was regarded as the ruler of the seas. Venus, in Roman mythology (identified with the Greek Aphrodite), was the goddess of love and beauty; she was born from the foam of the sea. Cupid, in Roman mythology (identified with the Greek Eros), was the god of love, represented as armed with bow and arrows which, when shot, became darts of desire.

In the older hierarchy of Greek mythology, Gaia (or Gaea), was the goddess of Earth, the wife of Ouranos and mother of the Titans, a race of giant gods and goddesses who were vanquished by the Olympian gods, who thus became the rulers of Olympus. Ouranos (in Roman mythology Uranus) was the son and husband of Gaia and father of the Titans: ruler of the firmament. He was overthrown by his son Kronos (Saturn in Roman mythology), who in turn was overthrown by Zeus (ruler of the presiding hierarchy of deities). Okeanos (more often Oceanus or simply Ocean), in Greek mythology a titan, the son of Ouranos and Gaia and father of the Oceanids (the sea nymphs), was represented as ruler of the sea, which was described as encircling the Earth. Tethys, the daughter of Ouranos and Gaia, was the wife of Okeanos and was regarded as a sea deity.

In Sanskrit terms, Aditi is a term used in connection with Infinite Space—to which *Ocean* (Okeanos) is equated in the quotation. Āditya, meaning the son of Aditi, is applied also to each one of the seven deities of the heavenly sphere, and especially to Varuna, who was regarded as the chief deity of the firmament. The other six deities are Mitra,

Aryaman, Bhaga, Daksha, Aṁśa, and Sūrya (the sun-god). Varuna is one of the oldest Vedic gods; regarded as the supreme deity and titled King of the Universe. He is described as fashioning and upholding heaven and earth and as possessing extraordinary power and wisdom. Though not generally associated with the ocean, he is connected with the waters of the atmosphere and the firmament. But in later mythology Varuna is indeed associated with waters and is regarded as the god of the ocean.

Mūlaprakriti signifies precosmic root substance—or primordial substance—that from which both Spirit (Purusha) and Matter (Prakriti) arise when Mūlaprakriti is activated (so to speak) by Fohat; in the quotation Fohat is equated with the Life-giving principle portrayed by the deity Eros.

The Separation of the Waters from the Land

It should be borne in mind that the ślokas of the Stanzas of Dzyan which have thus far been considered have dealt with a period of time prior to the peopling of the Earth and its condition. Although the next śloka, No. 11 in the series, is reminiscent of the creation story as related in the Bible, it tells also of the removal of the "turbid dark waters" from the land:

> The Lord of the Lords came. From her body he separated the waters, and that was heaven above, the first heaven (the atmosphere, or the air, the firmament).[12]

To the above śloka *The Secret Doctrine* adds this observation:

> Here tradition falls again into the Universal. As in the earliest version, repeated in the Purānas, so in the latest, the Mosaic account. In the first it is said: "He the Lord" (the god who has the form of Brahmā) "when the world had become one ocean concluding that within the waters lay the earth, and desirous to raise it up," to separate it, "created himself in another form." . . . In

12. Ibid. II:75; III:84, 6-vol. ed.; II:78, 3rd ed.

> the Elohistic "creation," "God" creates a firmament in the midst of the waters," and says "let *dry land* appear."[13]

The Elohistic version is narrated in Genesis. Here it is in full:

> And God said, Let there be a firmament in the midst of the waters, and let it divide the waters from the waters. And God made the firmament and divided the waters which were under the firmament from the waters which were above the firmament: and it was so. And God called the firmament Heaven. And the evening and the morning were the second day. And God said, Let the waters under the heaven be gathered together unto one place, and let the dry land appear; and it was so. And God called the dry land Earth; and the gathering together of the waters called the Seas: and God saw that it was good.[14]

The Hebrew Elohim are called in the translations "God."

> The Elohim are shown to "create"—that is to say, to build or to produce the two or "double heaven" (*not* Heaven and Earth); which means, in so many words, that they separated the upper manifested (angelic) heaven, or plane of consciousness, from the lower or terrestrial plane; the (to us) eternal and immutable Aeons from those periods that are in space, time and duration; Heaven from Earth, the unknown from the Known—to the profane.[15]

---

13. Ibid. II:75; III:85, 6-vol. ed.; II:78-9, 3rd ed. The account referred to as Purānic is quoted from the *Harivanśa* I, 36.

14. *Genesis,* I, 6-10.

15. SD II:488; IV:56-7, 6-vol. ed.; II:512-3, 3rd ed.

12.

HOW THE FORMS FOR PEOPLING THE EARTH WERE PROVIDED

THE STORY of the peopling of the Earth is resumed in śloka 12 of Stanza III. It tells how the appropriate forms were provided for the waiting monads. It should be borne in mind that the description of the Earth given here deals with a period of dormancy, when the human hosts had not yet entered on their cyclical evolution on Earth. In contrast to the narrative style of the previous ślokas, the twelfth is given in terse, forceful manner. It is as though commands were spoken:

> The great Chohans called the Lords of the Moon, of the airy bodies. "Bring forth men, men of your nature. Give them their forms within. She will build coverings without. Males-females will they be. Lords of the Flame also. . . ." (śloka 12)[1]

The meaning of the word "Chohan" has been amply explained in previous chapters. The Great Chohans who summon the Lunar Chohans are superior beings in evolutionary degree: their commands must be obeyed. H. P. Blavatsky asks the question: "Who are the Lords of the Moon?" and supplies the answer: "In India they are called *Pitris* or 'lunar ancestors.' "[2] They are sometimes also called "Lords of the Airy Bodies," signifying that their rūpas were superior to forms of material substance; they had graduated to a

---

1. SD II:16; III:29, 6-vol. ed.; II:17, 3rd ed.

2. Ibid. II:75; III:85, 6-vol. ed.; II:79, 3rd ed.

higher state, having evolved out of the inferior condition in which physical rūpas were used, to the next higher condition.

The Forms of the Lords of the Moon

"Bring forth men, men of your nature," was the command given to the Lords of the Moon. The vehicles which the Lords were to provide were definitely not of physical substance; certainly not the same as the bodies with which we are familiar today. A commentary from the Book of Dzyan specifically refers to the rūpas of the Lords of the Moon, although there they are mentioned as "inferior Lhas."

> It is from the material Worlds that descend they, who fashion physical man at the new Manvantaras. They are inferior Lhas (Spirits), possessed of a dual body (an astral within an ethereal form). They are the fashioners and creators of our body of illusion.[3]

The phrase "from the material Worlds" signifies that the Pitris have come from a more material chain of globes than the Earth Chain, because the fifth chain—in the embodiments of a planetary system—is not as material as the fourth, which is the most material of the series.[4]

The "new Manvantara" is the period of activity about to commence on the Earth (Globe D), following the completion of seven cycles of activity on Globe C. A comparable period of seven stages of evolutionary development takes place on each one of the seven globes of the Earth Chain. Furthermore, between the conclusion of seven cycles of activity on each one of the seven globes and the beginning of the periods of activity on the next globe, there are two periods of rest. The first rest-period which follows the seven cycles of activity is called a Sandhyā, or a Twilight. This is equal in length to one-tenth of the period of activity just concluded. The second rest-period—equal in length to the first—is termed a Sandhyānśa, or a Dawn, and it precedes the first new cycle of activity.

---

3. Ibid. II:57; III:67, 6-vol. ed.; II:60, 3rd ed.

4. This was explained in Chapter 10: "The Brothers of the Fifth."

The fact that the Lords of the Moon are termed "inferior Lhas" is significant. Lha is the Tibetan equivalent for the Sanskrit Dhyāni-Chohan. The Lords of the Moon are regarded as "inferior" because they were assigned the task of fashioning the rūpas. That is to say they were "inferior" to the Solar Pitris who, at a later time during the present Fourth Round on this Earth, were called upon to enlighten humanity. This was accomplished by activating Manas, the Mind-principle.

The rūpa (form) of the Lunar Pitris is described as a "dual body," not physical but ethereal, and an astral within the ethereal form. This astral within the ethereal describes what is termed the Linga-śarīra (enumerated as the second principle, counting upward, in the sevenfold constitution of man). This Sanskrit compound term signifies the model or pattern *(linga)* which is easily dissolved *(śarīra,* from the verbal root śrī, to dissolve).[5] As indicated, the "model body" would be "within" the outer vehicle termed the Sthūla-śarīra. However, in this description the rūpa was stated to be an ethereal, not a physical body. It was ethereal because of its having been evolved during the seven-round phase of evolutionary development on the Lunar Chain. Nevertheless, the densest phase of the material evolution had occurred during the fourth-round of the life-cycle on the Moon Chain.

The final phrase of the quotation from the Commentary, which describes the vehicle fashioned by the Lunar Pitris as "our body of illusion," refers to the present evolutionary stage on the Earth. It is termed a "body of illusion" because from birth it is continually subject to change—technically termed Nitya-Sarga and Nitya-Pralaya[6]—and finally "dies" when the immortal components leave the physical vehicle. These Sanskrit words represent one of the most important

5. The Linga-śarīra is also termed the "etheric double."

6. *Nitya* is a Sanskrit word meaning continual, perpetual, eternal; *sarga,* from the verbal root *srij,* signifies emission or creation of matter; *pralaya,* dissolution. These terms apply to the processes of incessant upbuilding and perpetual breaking down, which go on constantly in all aggregated bodies, whether man or planet.

concepts of the Esoteric Philosophy. They demonstrate the applicability of the Law of Periodicity, namely that for every period of activity there is a consequent period of rest: manifestation—represented by creation of forms—and disappearance, dissolution of forms. The process is continuous.

The Forms Provided By the Pitris

The Lunar Pitris were called upon by the great Chohans to provide rūpas (forms), because they had developed appropriate vehicles during the phase of humanity on the Moon Chain; they had produced vehicles suitable for that human phase of life. The command was given to them: "Bring forth men of your nature," signifying humans having rūpas of the kind the Lunar Lords had evolved for themselves after seven cycles of evolutionary development—"airy bodies," no less. However, the most significant point is this: *how* did the Lords of the Moon bring forth men? The clue is provided by the Stanzas, although it appears in the description of a later period, in Stanza VII, śloka 27. Therein reference is made to Kriyāśakti:

> It [the Third Race] created "Sons of Will and Yoga," by Kriyāśakti it created them.[7]

Kriyāśakti is defined as:

> . . . the mysterious *power of thought* which enables it to produce external, perceptible, phenomenal results by its own inherent energy. The ancients held that any idea will manifest itself *externally,* if one's attention (and Will) is deeply concentrated upon it; similarly, an intense volition will be followed by the desired result. . . .
>
> For creation is but the result of will acting on phenomenal matter, the calling forth out of it the primordial divine *Light* and eternal *Life.*[8]

Since the Lords of the Moon had accomplished seven rounds of evolutionary development, they would have gained

---

7. SD II:19; III:31, 6-vol. ed.; II:21, 3rd ed.

8. Ibid. II.173; III:180, 6-vol. ed.; II:182-3, 3rd ed.

knowledge of, and proficiency in, the use of Kriyāśakti. For even on this Earth, creation by Kriyāśakti was accomplished, as related in the Stanzas.

> The Pitris shoot out from their ethereal bodies, still more ethereal and shadowy similitudes of themselves, or what we should now call "doubles," or "astral forms," in their own likeness. This furnishes the Monad with its first dwelling, and blind matter with a model around and upon which to build henceforth.[9]

The rūpas that were provided were not physical, as śloka 12 goes on to explain: "Give them (i.e., the Jīvas or Monads) their forms within." (The words jīva and monad are used synonymously in the Stanzas, as already indicated). It should be understood that the Lords of the Moon are not to be regarded as the monads who will be accomplishing the *human* circuit of the Fourth Round; they are providing the rūpas for the monads. These rūpas, however, are not the *external* bodies. The Lords of the Moon are commanded to provide "internal rūpas" similar to their own. It is this internal rūpa which is termed the Linga-śarīra.

The fourth sentence of the śloka explains how the external "coverings" for the monads will be provided: "She (Mother Earth or Nature) will build coverings without." Then, a Commentary was added to a later śloka, one portion of which is applicable at this point:

> They were the shadows of the shadows of the Lords. They (the shadows) expanded. The Spirits of the Earth clothed them.[10]

The first word of the Commentary, "they," refers to the rūpas which were produced by the Lords of the Moon. The first sentence signifies that "the progenitors created man out of their own astral bodies."[11] The Sanskrit word for shadows is Chhāyās.

---

9. Ibid. I:248; I:293, 6-vol. ed.; I:268, 3rd ed.

10. Ibid. II:109; III:118, 6-vol. ed.; II:115, 3rd ed.

11. Ibid. II:112; III:121, 6-vol. ed.; II:118, 3rd ed.

In connection with the second sentence of the Commentary, the following explanation is given:

> To these "Spirits" belong temporarily the human astral selves: and it is they who give, or build, the physical tabernacle of man, for the Monad and its conscious principle, Manas, to dwell in.[12]

Furthermore, it is the "Spirits of the Earth" or the elementals of the three lowest kingdoms (lower than the manifested kingdoms of Nature) who provide the external bodies, the "clothing."

In one of the first series of the Stanzas of Dzyan, the question is asked: "Who forms man?" and the answer is given: "The seven lives and the One Life."[13] The One Life refers to the origin of man's immortal essence, an emanation from the Supreme, or Paramātman. A Commentary from the Book of Dzyan is quoted:

> Every visible thing in this Universe was built by such LIVES, from conscious and divine primordial man down to the unconscious agents that construct matter. . . . From the ONE LIFE formless and Uncreate, proceeds the Universe of lives.[14]

The seven lives signify the seven kinds of life-atoms pertaining to the seven kingdoms. The seven kingdoms are: the three elemental kingdoms, the mineral kingdom, the plant kingdom, the animal kingdom, and the human kingdom. A life-atom may be defined as the ensouling power in a physical atom. *The Secret Doctrine* explains:

> The Occultists trace every atom in the universe, whether an aggregate or single, to One Unity, or Universal *Life*. . . . We know and speak of "life-atoms"—and of "sleeping-atoms"—because we regard these two forms of energy—the kinetic and the potential—as produced by one and the same force or the ONE LIFE, and regard

---

12. Ibid. II:110; III:119, 6-vol. ed.; II:116, 3rd ed.

13. Ibid. Stanza VII, śloka 5 (first series), I:34; I:98, 6-vol. ed.; I:66, 3rd ed.

14. Ibid. I:250; I:294, 6-vol. ed.; I:270, 3rd ed.

the latter as the source and mover of all.[15]

The Occult doctrine says: Not only the chemical compounds are the same, but the same infinitesimal *invisible lives* compose the atoms of the bodies of the mountain and the daisy, of man and the ant, of the elephant, and of the tree which shelters him from the sun. Each particle—whether you call it organic or inorganic—*is a life.* Every atom and molecule in the Universe is both *life-giving* and *death-giving* to that form, inasmuch as it builds by aggregation universes and the ephemeral vehicles ready to receive the transmigrating soul, and as eternally destroys and changes the *forms* and expels those souls from their temporary abodes. It creates and kills; it is self-generating and self-destroying; it brings into being, and annihilates, that mystery of mysteries—the *living body* of man, animal, or plant, every second in time and space; and it generates equally life and death, beauty and ugliness, good and bad, and even the agreeable and disagreeable, the beneficent and maleficent sensations. It is that mysterious LIFE, represented collectively by countless myriads of lives.[16]

The Pitris Provide an Androgynous Form

The consideration of the twelfth śloka is now resumed. The next sentence is a remarkable one: "Males-females will they be." This would seem to imply that the Lords of the Moon were androgynous; for evidently they had attained such an evolutionary status. But this state did not take place for humanity until some epochs later—not until their "coverings" had been completed. However, the Third Root-Race is described as having had an androgynous form during some periods of its sojourn on earth. *The Secret Doctrine* explains:

> Almost sexless, in its early beginnings, it became bisexual or androgynous; very gradually of course. The passage from the former to the latter transformation

15. Ibid. II:672; IV:241-2, 6-vol. ed.; II:709-10, 3rd ed.

16. Ibid. I:261; I:305, 6-vol. ed.; I:281-2, 3rd ed.

> required numberless generations, during which the simple cell that issued from the earliest parent (the two in one), first developed into a bisexual being; and then the cell, becoming a regular egg, gave forth a unisexual creature. The Third Race mankind is the most mysterious of all the hitherto developed five Races.[17]

Our present humanity is in its fifth major evolutionary developmental stage, also described as the Fifth Root-Race, in the Fourth Round on this Earth.

This passage from *Isis Unveiled* relative to androgynous humanity is worthy of note, explaining as it does the biblical allegory:

> Starting as a pure and perfect spiritual being, the Adam of the second chapter of *Genesis,* not satisfied with the position allotted to him by the Demiurgus (who is the eldest first-begotten, the Adam-Kadmon), Adam the second, the "man of dust," strives in his pride to become Creator in his turn. Evolved out of the androgynous Kadmon, this Adam is himself an androgyn; for, according to the oldest beliefs presented allegorically in Plato's *Timaeus,* the prototypes of our races were all enclosed in the microcosmic tree which grew and developed within and under the great mundane or macrocosmic tree. Divine spirit being considered a unity, however numerous the rays of the great spiritual sun, man has still had his origin like all other forms, whether organic or otherwise, in this one Fount of Eternal Light. Were we even to reject the hypothesis of an androgynous man, in connection with physical evolution, the significance of the allegory in its spiritual sense, would remain unimpaired. So long as the first god-man, symbolizing the two first principles of creation, the dual male and female element, had no thought of good and evil he could not hypostasize "woman," for she was in him as he was in her. It was only when, as a result of the evil hints of the serpent, *matter,* the latter condensed itself and cooled on the spiritual man in its contact with the elements, that the fruits of the man-tree—who

17. Ibid. II:197; III:202, 6-vol. ed.; II:207, 3rd ed.

> is himself that tree of knowledge—appeared to his view. From this moment the androgynal union ceased, man evolved out of himself the woman as a separate entity. They have broken the thread between pure spirit and pure matter. Henceforth they will create no more *spiritually,* and by the sole power of their will; man has become a physical creator, and the kingdom of spirit can be won only by a long imprisonment in matter.[18]

A similar reference to the duality of Spirit and Matter is found in Hindu scriptures, where Purusha (Spirit) is described as mounting on Prakriti's (Matter's) shoulders. The interpretation of this symbology is supplied, as well as the significance of the potential androgyne separating into the two sexes.

> Besides the material which will be needed for its future human form, the monad requires (*a*) a spiritual model, or prototype, for that material to shape itself into; and (*b*) an intelligent consciousness to guide its evolution and progress, neither of which is possessed by the homogeneous monad, or by senseless though living matter. The Adam of dust requires the *Soul of Life* to be breathed into him: the two middle principles, which are the *sentient* life of the irrational animal and the Human Soul, for the former is irrational without the latter. It is only when, from a potential androgyne, man has become separated into male and female, that he will be endowed with this conscious, rational, individual Soul (*Manas*) "the principle, or the intelligence, of the Elohim," to receive which, he has to eat of the fruit of Knowledge from the Tree of Good and Evil. How is he to obtain all this? The Occult doctrine teaches that while the monad is cycling on downward into matter, these very Elohim—or Pitris, the lower Dhyān-Chohans—are evolving *pari passu* with it on a higher and more spiritual plane, descending also relatively into matter on their own plane of consciousness, when, after having reached a certain point, they will meet the incarnating senseless monad, encased in the

---

18. *Isis Unveiled,* I:297.

> lowest matter, and blending the two potencies, Spirit and Matter, the union will produce that terrestrial symbol of the "Heavenly Man" in space—PERFECT MAN. In the Sānkhya philosophy, Purusha (spirit) is spoken of as something impotent unless he mounts on the shoulders of Prakriti (matter), which, left alone, is—senseless. But in the secret philosophy they are viewed as graduated. Though one and the same thing in their origin, Spirit and Matter, when once they are on the plane of differentiation, begin each of them their evolutionary progress in contrary directions—Spirit falling gradually into matter, and the latter ascending to its original condition, that of a pure spiritual substance. Both are inseparable, yet ever separated. In polarity, on the physical plane, two like poles will always repel each other, while the negative and the positive are mutually attracted, so do Spirit and Matter stand to each other—the two poles of the same homogeneous substance, the root-principle of the universe.[19]

The root-principle of the universe is also known as Mūlaprakriti (usually referred to as precosmic root-substance), ever-present in the universe but solely in an unmanifested stage. Its first phase of manifestation is termed Purusha-Prakriti, Spirit-Matter—which are "one and the same thing in their origin," prior to differentiation. As the stages of the manifestation of a system proceed, a separation of Spirit and Matter takes place as the cyclic phases unfold. When the plane of differentiation has been entered Spirit and Matter pursue their evolutionary progress in contrary directions; this is referred to as the evolution of matter and the involution of spirit on the descending arc of the evolutionary cycle. Then on the ascending arc of the evolutionary cycle the evolution of spirit and the involution of matter takes place.

There is yet to consider the closing sentence of the twelfth śloka, which is indeed perplexing. The ellipsis (or dots) followed the word "also" in the original edition, giving the impression that the sentence was unfinished: "Lords of the Flame, also. . . ."

---

19. SD I:247-8; I:292-3, 6-vol. ed.; I:267-8, 3rd ed.

Does the sentence mean that the Lords of the Flame also were summoned? Or that they had achieved their evolutionary status on the Moon Chain? Or does it signify that the jīvas or monads entering the Earth will be "Lords of the Flame" eventually? Or that the Lords of the Moon, "of the airy bodies" will be "Lords of the Flame, also?"[20]

20. As śloka 13 of Stanza III also deals with the Lords of the Flame, further explanation will be provided in the exposition of that śloka in the next chapter.

13.

THE LORDS OF THE FLAME

"FIRE" or "Flame" is commonly used in esoteric literature as a symbol for the mind-principle, Manas. The concept is present also in Greek mythology in the story of Prometheus, who is described as bringing fire to mankind from celestial regions.

The Lords of the Flame are first mentioned in connection with the peopling of the Earth in śloka 12 of the Third Stanza of Dzyan, but the assistance they rendered to humanity is not described until later stanzas. In this connection they are called by many names, and descriptive epithets are applied to them. A distinction should be made between the two groups of Pitris: the inferior and the superior. The inferior Pitris are the Lords of the Moon, who produced the *forms* for peopling the Earth. The superior Pitris are the Lords of the Flame and are grouped under three Sanskrit terms: Mānasaputras, Agnishvāttas, Kumāras. Three aspects are thus stressed: (1) the Manas aspect; (2) the Agnishvātta aspect; (3) the Kumāra aspect.

(1) The Manas aspect deals with the awakening of the Mind principle. Consequently the name Mānasaputras, which means the Sons of Mind, is particularly applicable to the phase of the enlightenment of mankind.

(2) The Agnishvātta aspect stresses the evolutionary phase, with special reference to the successful completion of the development of Manas, for these Solar Pitris are completely developed in this respect. The first component of the Sanskrit term, Agni, means fire; here again the word is associated with the principle of the mind, Manas.

(3) The Kumāra aspect signifies the enlightenment of infant humanity by direct assistance, in that the Holy Yogins (as the Kumāras are termed) actually incarnated in a portion of the human race. Although this function of the Kūmaras is the esoteric idea, the exoteric version is different and will be considered later.

Here are the descriptive names applied to the Lords of the Flame:

Solar Pitris
Solar Fathers
Solar Ancestors
Solar Angels
Solar Devas
Higher Angels
Gods of Will
Mānasaputras
Mānasas
Mānasvin
Mānasa Devas
Mānasa Dhyānis
Dhyānis
Spiritual Dhyānis
Sons of Manas

Sons of Mind
Sons of Wisdom
Lords of Wisdom
Lords of Spiritual Life
Sons of Mahat
Mind-born Sons
Mānasa Pitris
Lords of the Flame
Agnishvāttas
Agnishvātta Pitris
Sons of Fire
Fire Dhyānis

Fiery Egos
Kumāras
Holy Yogins
Virgin Youths
Virgin Ascetics
Vairājas
Holy Youths
Rājasas
Abhūtarajasas
Asuras
Arūpa Pitris
Incorporeal Pitris
Three Higher Classes of Pitris
Celestial Ancestors

The Names of the Lords of the Flame Described

SOLAR PITRIS. Two distinct groups of Pitris assisted in the evolutionary development of humanity; one, the Lunar Pitris, provided mankind with its corporeal vehicle, the outer man; the other, the Solar Pitris, enlightened the inner man by awakening the mind-principle, Manas.

SOLAR FATHERS. The word "fathers" is a literal translation of the Sanskrit word *pitris.* "Solar" is used to distinguish between those who assisted in the evolutionary development of the inner man and those called the Lunar Fathers, who projected the outer form of man.

SOLAR ANCESTORS. This term directs attention to the link which is present between this group of celestial beings and mankind. ("Celestial" here signifies beings who have evolved from the human kingdom in past manvantaras.) "They are in fact more truly our *Pitris* (ancestors) than the Pitris who projected the first physical men,"[1] i.e. the Lunar Pitris.

SOLAR ANGELS. An equivalent term for Solar Pitris. A passage which describes them, also clearly indicates their function:

> The Endowers of man with his conscious, immortal EGO, are the "Solar Angels"—whether so regarded metaphorically or literally. The mysteries of the Conscious EGO or human Soul are great. The esoteric name of these "Solar Angels" is, literally, the "Lords" (*Nāthas*) of persevering ceaseless devotion (pranidhāna).[2]

SOLAR DEVAS. The Sanskrit word *deva* means a "shining being." *Deva* is derived from the verbal root *div,* to shine. "Solar" here assigns the devas to the Solar Pitris.

HIGHER ANGELS. Equivalent to Solar Angels (q.v.). The following is a passage from *The Divine Pymander,* telling of the acquisition of the sacred fire. It is reminiscent of the story of Prometheus:

> The higher Angels had broken, countless aeons before, through the "Seven Circles," and thus *robbed* them of the Sacred fire; which means in plain words, that they had assimilated during their past incarnations, in lower as well as in higher worlds, all the wisdom therefrom—the reflection of MAHAT in its various degrees of intensity.[3]

Mahat signifies Universal Mind.

GODS OF WILL. This term is used in a Commentary on the Book of Dzyan, and indicates that these celestial beings descended to Earth in order to accomplish their purpose: "It is those who have deserted the Superior Spheres, the Gods of Will, who complete the Manu of illusion."[4] "The Manu

---

1. SD, I:219 fn.; I:266, 6-vol. ed.; I:239, 3rd ed.
2. Ibid. II:88; III:97, 6-vol. ed.; II:92, 3rd ed.
3. Ibid. II:80-1; III:90, 6-vol. ed.; II:84, 3rd ed.
4. Ibid. II:57; III:68, 6-vol. ed.; II:60, 3rd ed.

of illusion" is representative of the condition of man before the Solar Pitris awakened the mind-principle, Manas.

MĀNASAPUTRAS. A Sanskrit compound of *mānasa,* the adjectival form of *manas,* the mind, and *putras,* sons; hence the class of celestial beings who enlighten humanity by enabling human beings to use the mind-principle.

MĀNASAS. A class of celestial beings who are associated with humanity. They "incarnated in the Third Race, and in this and various other ways endowed mankind with Mind."[5] MĀNASAM is the accusative form of the word Manas.

MĀNASVIN. The adjectival form, derived from the verbal root *man,* to think; hence full of mind, wise. The term is used to designate "all those high celestial beings whom Occult teaching calls Mānasvin, the Wise."[6]

MĀNASA DEVAS. The Sanskrit compound signifies the "shining beings" who awaken the mind principle. It is "the Mānasa Devas who have endowed man with the consciousness of his immortal soul."[7]

MĀNASA DHYĀNIS. This compound refers to the celestial beings who represent the mānasic, or intelligence aspect, in the Dhyāni-Chohanic Kingdom. Because of their having undergone their evolutionary development on a planetary system prior to the Lunar Chain, as well as on the Moon Chain itself, before coming into manifestation on the Earth Chain, the Mānasa Dhyānis were able to help humanity, here referred to as Lunar Monads:

> The "Lunar Monads" cannot progress, for they have not yet had sufficient touch with the forms created by "Nature" to allow of their accumulating experiences through its means. It is the Mānasa-Dhyānis who fill up the gap, and they represent the evolutionary power of Intelligence and Mind, the link between "Spirit" and "Matter"—in this Round [the Fourth Round].[8]

---

5. Ibid. II:89; III:98, 6-vol. ed.; II:93, 3rd ed.

6. Ibid. II:90; III:99, 6-vol. ed.; II:95, 3rd ed.

7. Ibid. II:525; IV:94, 6-vol. ed.; II:552, 3rd ed.

8. Ibid. I:181-2; I:233, 6-vol. ed.; I:204, 3rd ed.

DHYĀNIS. The word is derived from the Sanskrit verbal root *dhyai,* to meditate. It signifies one who meditates. It is used in *The Secret Doctrine,* as well as in the Stanzas and Commentaries, as an abbreviation for Dhyāni-Chohans, meaning celestial beings who act as Regents. Therefore, one must take care to decide whether the term is applicable to Solar Pitris or Lunar Pitris.

SPIRITUAL DHYĀNIS. An equivalent for Lords of the Flame.

SONS OF MANAS. Manas is the fifth principle (counting upward) in the sevenfold constitution of man—the mind-principle. So this is equivalent to the Sanskrit term Mānasaputras; *putras*—sons.

SONS OF MIND. A literal translation of the Sanskrit compound term Mānasaputras.

SONS OF WISDOM. An equivalent to Mānasaputras or Lords of the Flame. The term is first used in Stanza VII, śloka 24: "The Sons of Wisdom, the Sons of Night (issued from the body of Brahmā when it became Night), ready for re-birth, came down." The Sons of Night who issued from the body of Brahmā signifies (in one interpretation) that the Sons of Wisdom were active in a previous Manvantara: they "came down" into this sphere from a superior loka. Śloka 25 continues:

> How did the Mānasa, the Sons of Wisdom act? They rejected the Self-born (the boneless). They are not ready. They spurned the (First) Sweat-Born. They are not quite ready. They would not enter the (First) Egg-Born.[9]

The Secret Doctrine explains:

> To a Theist or a Christian this verse would suggest a rather theological idea: that of the Fall of the Angels through Pride. In the Secret Doctrine, however, the reasons for the refusal to incarnate in *half-ready* physical bodies seem to be more connected with physiological than metaphysical reasons. Not all the organisms were sufficiently ready. The incarnating powers chose the

9. Ibid. II:171; III:178, 6-vol. ed.; II:180, 3rd ed.

ripest fruits and spurned the rest.[10]

Here is another passage showing that the Mānasaputras had attained their status in previous Manvantaras, and because of that factor were able to assist humanity in awakening the mind principle.

> The Sons of Wisdom, or the *spiritual* Dhyānis, had become "intellectual" through their contact with matter, because during previous cycles of incarnation, they had already reached that degree of intellect which enabled them to become independent and self-conscious entities, *on this plane* of matter. They were reborn only by reason of Karmic effects. They *entered* those who were "ready," and became the Arhats, or sages.[11]

Lords of Wisdom. This term indicates that the celestial beings who assisted in the evolutionary development of humanity had activated Manas, the mind principle, in its highest aspect: that of the power of true and right discernment. The term is first used in Stanza VII, śloka 26, wherein the stages of evolution taking place in the Third Root-Race are described, after the Second Root-Race had passed through its phase of procreation termed the Sweat-born.

> When the Sweat-born produced the Egg-born, the Two-fold (androgyne Third Race), the Mighty, the Powerful with Bones, the Lords of Wisdom said: "Now shall we create."

Why "now"—and not earlier? This the following śloka explains:

> (Then) the Third (race) became the vāhana (vehicle) of the Lords of Wisdom. It created Sons of "Will and Yoga," by Kriyāśakti, it created them, the Holy Fathers, Ancestors of the Arhats. (śloka 27)

The Secret Doctrine adds this explanation:

> How did they *create*, since the "Lords of Wisdom" are identical with the Hindu Devas, who refuse "to create"?

10. Ibid.

11. Ibid. II:167; III:174, 6-vol. ed.; II:176, 3rd ed.

> Clearly they are the *Kumāras* of the Hindu Pantheon and Purānas, those elder sons of Brahmā, "Sanandana and the other sons of *Vedhas*," who, previously created by him "without desire or passion, remained chaste, full of holy wisdom and undesirous of progeny"? (*Vishnu-Purāna,* Bk. I, ch. 7)
>
> The power, by which they first created, is . . . Kriyā-śakti—the mysterious *power of thought* which enables it to produce external, perceptible, phenomenal results by its own inherent energy.[12]

Sanandana is the name of one of the seven mind-born sons of Brahmā. *Vedhas* means worshippers of the gods.

The Lords of Wisdom are thus made equivalent to the Kumāras—one of the three main classes into which the Solar Pitris have been designated.

LORDS OF SPIRITUAL LIFE. A term used in a Commentary from the Book of Dzyan, referring to the spiritualizing aspect which the great Lords were able to perform, thereby assisting the evolution of humanity. They are likened to the Sons of Mahat.

> The Sons of Mahat are the quickeners of the human Plant. They are the Waters falling upon the arid soil of latent life, and the Spark that vivifies the human Animal. They are the Lords of Spiritual Life Eternal. . . . In the beginning (in the Second Race) some (of the Lords) only breathed of their essence into Mānushya (men), and some took in man their abode.[13]

SONS OF MAHAT. The Sanskrit term Mahat is usually rendered "Universal Mind." However, in a Commentary from the Book of Dzyan the Sons of Mahat are described as though they are equivalent to the Mānasaputras, thus conveying the idea that the Mānasaputras are the efflux of the Divine Mind:

> The Sons of Mahat are the quickeners of the human Plant. They are the Waters falling upon the arid soil

12. Ibid. II:172-3; III:179-80, 6-vol. ed.; II:182, 3rd ed.

13. Ibid. II:103; III:111, 6-vol. ed.; II:108, 3rd ed.

of latent life, and the Spark that vivifies the human animal. They are the Lords of Spiritual Life eternal.[14]

MIND-BORN SONS. The significance of this term is given in a remarkable passage:

> It [devotion] lives undeniably, and has settled in all its ineradicable strength and power in the Asiatic Aryan heart from the Third Race direct through its first "Mind-born" sons—the fruits of *Kriyāśakti*.[15]

MĀNASA PITRIS. The two Sanskrit words literally mean the Fathers of the Mind. The Solar Pitris are so regarded because they awakened the mind-principle in humanity during the Third Root Race.

LORDS OF THE FLAME. A symbolical name for the Solar Pitris, used in the Stanzas of Dzyan. In *The Secret Doctrine* the question is asked: Who are the Lords of the Flame? and the reply is:

> These are then those "Flames" (the *Agnishvātta*) who, as shown in Śloka 13, "remain behind" instead of going along with the others to create men on Earth. But the true esoteric meaning is that most of them were destined to incarnate as the *Egos* of the forthcoming crop of Mankind. . . . If the reader were told, as in the *semi-esoteric* allegories, that these Beings were returning *Nirvānīs,* from preceding *Mahā-Manvantaras* . . . he would hardly understand the text correctly.[16]

AGNISHVĀTTA. This is a compound Sanskrit word, formed of *agni,* fire, or inner essence, applied figuratively to the mind, hence "the fire of mind"; *śvatta,* is the past participle form of the verbal root *svad* meaning to taste or to sweeten. Thus the term may be explained in two ways: (1) as applied to those Beings who have tasted of the fire of mind and have become stimulated thereby to achieve and conclude their cycle of evolution so as to graduate into a higher Class; (2) as meaning those Beings who have been sweetened by the

---

14. Ibid. II:103; III:111, 6-vol. ed.; II:108, 3rd ed.

15. Ibid. I:211; I:259, 6-vol. ed.; I:232, 3rd ed.

16. Ibid. II:79; III:89, 6-vol. ed.; II:83, 3rd ed.

fire of suffering and experience inseparable from passing through the Cycle of Existence, or the Circle of Necessity, and have thereby achieved graduation.

Agnishvātta Pitris. By the addition of the Sanskrit word *Pitris,* meaning fathers, attention is directed to the role played by the Agnishvāttas as fathers to "infant humanity"—termed "infant" because unawakened; the mind-principle was not yet functioning. "The *Vāyu-Purāna* shows the region called Virāja-loka inhabited by the Agnishvāttas."[17] Virāja-loka is another name for Janar-loka—one of the lokas superior to Bhūr-loka (our Earth).

Sons of Fire. Another term for the Agnishvāttas. Bearing in mind the significance of the symbol of fire, the following quotation explains why the Sons of Fire are so called:

> . . . because they are the first Beings (in the Secret Doctrine they are called "Minds"), evolved from Primordial Fire.[18]

In the first series of the Stanzas of Dzyan, the Solar Pitris are called the Sons of Fire:

> Listen, ye Sons of the Earth, to your instructors—the Sons of the Fire. Learn there is neither first nor last; for all is one number, issued from no number. [Stanza IV, śloka 1]
>
> These terms, the "Sons of the Fire," the "Sons of the Fire-Mist," and the like, require explanation. . . .
>
> With these verses the mystic sense of the solar and lunar symbols is connected: the Pitris are *lunar* deities and our ancestors, because they *created the physical man.* The Agnishvāttas, the Kumāras (the seven mystic sages), are solar deities, though the former are Pitris also; and these are the "fashioners of the *Inner* Man."
>
> "The fire or knowledge burns up all action on the plane of illusion," says the commentary. "Therefore, those who have acquired it and are emancipated, are called "Fires."[19]

---

17. Ibid. II 89 fn.; III:98, 6-vol. ed.; II:94, 3rd ed.

18. Ibid. I:87; I:151, 6-vol. ed.; I:114, 3rd ed.

19. Ibid. I:86-7; I:150-1, 6-vol. ed.; I:113-5, 3rd ed.

FIRE DHYĀNIS. Another term for the Agnishvāttas.

> Now, with regard to the seven classes of Pitris, each of which is again divided into seven, a word to students. . . . That class of the "Fire Dhyānis," which we identify on undeniable grounds with the Agnishvāttas, is called in our school the "Heart" of the Dhyān-Chohanic Body; and it is said to have incarnated in the third race of men and made them perfect. The esoteric Mystagogy speaks of the mysterious relation existing between the hebdomadic essence or substance of this angelic Heart and that of man.[20]

FIERY EGOS. In this term allusion is made to the symbol of fire with reference to the Manas principle centered in the Ego. The "Fiery Egos" are allocated to the Vairājas (q.v.) as belonging to other Manvantaras.

KUMĀRAS. Usually rendered Virgin Youths. This Sanskrit word is defined in two ways: (1) compounded of *ku,* easily, and *māra,* dying, from the verbal root *mri,* to die; (2) derived from the verbal root *kam,* to wish or long for.

> The Kumāras are, exoterically, "the creation of Rudra or Nilalohita, a form of Śiva, by Brahmā, and of certain other mind-born sons of Brahmā." But, in the esoteric teaching, they are the progenitors of the true spiritual SELF in the physical man—the higher Prajāpati, while the Pitris, or lower Prajāpati, are no more than the *fathers* of the model, or type of his physical form, made "in *their* image."[21]
>
> They [the Kumāras] may indeed mark a "special" or extra *creation,* since it is they who, by incarnating themselves within the senseless human shells of the two first Root-races, and a great portion of the Third Root-race—create, so to speak, a *new race*: that of thinking, self-conscious and *divine* men.[22]

Seven Kumāras are named, four in exoteric works: Sanat-Kumāra, Sananda, Sanaka, Sanātana; three others are added

---

20. Ibid. II:91; III:100, 6-vol. ed.; II:96, 3rd ed.

21. Ibid. I:457; II:176, 6-vol. ed.; I:493, 3rd ed.

22. Ibid.

in esoteric lore: Sana, Kapila, and Sanatsujāta.

HOLY YOGINS. In exoteric literature, such as the Purānas, the Kumāras are always described as Yogins.

VIRGIN YOUTHS. Since one of the meanings ascribed to the Sanskrit word Kumāra is youth, Hindu exoteric works represent the Kumāras as Yogins, whose piety inspired them to refuse to create progeny because of wishing to remain Virgin Youths so that they might attain final liberation. "But agreeably to esoteric interpretation, it was a self-sacrifice for the benefit of mankind," adds *The Secret Doctrine*.[23]

VIRGIN ASCETICS. In another Hindu exoteric account the Kumāras are described as Virgin Ascetics who refuse to create the material being, man.

VAIRĀJAS. One of the ancient Sanskrit works, the *Harivanśa* regards the Vairājas as

> . . . one of the classes of Pitris—a statement corroborated in the Secret Teachings, which, however, identify the Vairājas with the *elder Agnishvāttas* and the *Rājasas,* or *Abhūtarajasas,* who are incorporeal without even an astral phantom.[24]
>
> The Vairājas belong to, are the fiery Egos of, other Manvantaras. They have already been purified in the fire of passions. It is they who refused to create. They have reached the Seventh Portal, and have refused Nirvāna, remaining for succeeding Manvantaras.[25]

HOLY YOUTHS. A term applied to the Vairājas, of whom it is related in Stanza VI, śloka 5 of the first series:

> At the fourth (Round, or revolution of life and being around "the seven smaller wheels") the Sons are told to create their images. One third refuses. Two (thirds) obey. . . .

Says the Commentary explaining the verse:

> "The holy youths (the gods) refused to multiply and create species after their likeness, after their kind. They

---

23. Ibid. II:243; III:245, 6-vol. ed.; II:253, 3rd ed.

24. Ibid. II:89; III:98, 6-vol. ed.; II:93-4, 3rd ed.

25. Ibid. V:543, 6-vol. ed.

are not fit forms (rūpas) for us. They have to grow. They refuse to enter the chhāyās (shadows or images) of their inferiors."[26]

RĀJASAS. The Sanskirt lexicon defines this word as a class of gods in the fifth manvantara. One of the classes of Solar Pitris, who are associated with the Vairājas.

ABHŪTARAJASAS. The name of a class of deities of the fifth manvantara. Associated with the Vairājas. (q.v.)

ASURAS. Another class of Solar Pitris, portrayed in a dual capacity; even the word is given two meanings: (1) asura, from *asu,* a breath, a life, hence a divine being, represented in the Vedas as a spiritual being, a god. (2) Compounded of *a*—not; *sura,* a god; hence "not a god"; therefore in later literature the Asuras are described in the Purānas as fallen deities, demons. This is clearly pointed out in a Commentary:

> The subtle bodies remain without understanding (Manas) until the advent of the Suras (Gods) now called Asuras (not Gods).
>
> "*Not-gods,*" for the Brahmans, perhaps, but the highest *Breaths,* for the Occultist; since those Progenitors (*Pitaras*), the formless and the intellectual, refuse to build man, but endow him with mind.[27]

ARŪPA PITRIS. Another name for the Solar Pitris. The Sanskrit word *arūpa* is a compound of *a,* without; *rūpa* form; hence formless.

> The Pitris being divided into *seven classes,* we have here the mystic number again. Nearly all the Purānas agree that three of these are *arūpa,* formless, while four are corporeal; the former being intellectual and spiritual, the latter material and devoid of intellect.[28]
>
> Esoteric philosophy, however, teaches that *one third* of the Dhyānis—i.e., the three classes of the *Arūpa* Pitris, endowed with intelligence, "which is a formless breath, composed of *intellectual* not elementary substances" (*Harivanśa,* 932)—was simply doomed by the law of

26. Ibid. I:191-2; I:242-3, 6-vol. ed.; I:213-4, 3rd ed.

27. Ibid. II:92; III:101, 6-vol. ed.; II:96-7, 3rd ed.

28. Ibid. II:91; III:100, 6-vol. ed.; II:96, 3rd ed.

Karma and evolution to be reborn (or incarnated) on Earth.[29]

INCORPOREAL PITRIS. Synonymous with Arūpa Pitris (q.v.). The three higher classes of Pitris—the Solar Pitris—were without physical forms, as explained above. The Corporeal Pitris are termed the Lunar Pitris or Barhishad Pitris. (q.v.).

THREE HIGHER CLASSES OF PITRIS. Of the seven classes of Pitris, the three higher, or incorporeal, are the Solar Pitris. As well as the many names already explained, the exoteric works mention still others: Ajitas, Tushitas, Satyas, Haris, Vaikunthas, Sādhyas, and Ādityas. "But they are identical with the Mānasa or Rājasas, and these with our incarnating Dhyān-Chohans."[30]

CELESTIAL ANCESTORS. In this term, reference is made to the beings who are superior to the human kingdom, hence regarded as "celestial." Then, too, the link with humanity is stressed by the word "ancestors." For *The Secret Doctrine* refers to the Celestial Ancestors as "Entities from preceding worlds" who "step in on this our plane."[31] This is important for two reasons: (1) It calls attention to the fact that these suprahuman beings had achieved their evolutionary status not only on *worlds preceding the earth,* but also prior to the lunar system; thus they are superior to the Lunar Pitris. (2) Because of having evolved in preceding worlds of which the present Earth planetary system is a reimbodiment, therefore these Celestial Ancestors have an attachment to the Earth which is prior to any of the other planetary systems of the solar system (this includes the planet Venus), because the present planets are more recent than the preceding worlds in which the Celestial Ancestors functioned.

This review of the forty-one names of the Solar Pitris shows well the significance of the Lords of the Flame in the evolution of the monads during their cycle of evolution in the human kingdom. However, it should be understood that their actual assistance was not provided during the

29. Ibid. II:93; III:102, 6-vol. ed.; II:98, 3rd ed.

30. Ibid. II:90; III:99, 6-vol. ed.; II:94, 3rd ed.

31. Ibid. I:248; I:293, 6-vol. ed.; I:268, 3rd ed.

particular period described in śloka 13, for that śloka deals with an epoch prior to the time when the earth was peopled.

The Activities of the Pitris

In the narrative of the peopling of the Earth as presented in the Stanzas of Dzyan, śloka No. 13 tells of the activities of the Pitris, both Lunar and Solar:

> They went each on his allotted land: seven of them each on his lot. The Lords of the Flame remain behind. They would not go, they would not create.[32]

The first word of the śloka, "they," specifies the Lunar Pitris, or the Moon-Gods.[33] But the second and third sentences apply to the Solar Pitris. "The Lords of the Flame remain behind." This was the subject of Chapter 10, entitled "No fit rūpas for the Brothers of the Fifth," and was fully considered therein. As for the third sentence, the exoteric account is given in the Hindu scriptures, as for instance in the *Vishnu-Purāna*:[34]

> The allegory says that Sanandana and other *Vedhas,* the Sons of Brahmā, his first progeny, "were without desire or passion, inspired with the holy wisdom, estranged from the Universe and *undesirous of progeny.*"[35]

But the esoteric significance of the sentence "they would not create" is given in the words of a Commentary:

> The primordial Emanations from the creative Power are too near the absolute Cause. They are transitional and latent forces, which will develop only in the next and subsequent removes.[36]

Then there is another significant passage which opens with a question to which the answer is supplied:

---

32. Ibid. II:16; III:29, 6-vol. ed.; II:18, 3rd ed.

33. Further exposition of this sentence of śloka 13 is provided in Chapter 14.

34. Wilson's translation, Vol. I:101-2.

35. SD II:78; III:87, 6-vol. ed.; II:81-2, 3rd ed.

36. Ibid.

> But who are "They" who create, and the "Lords of the Flame," "who do not"? Occultism divides the "Creators" into twelve classes, of which four have reached *liberation* to the end of the "Great Age," the fifth is ready to reach it, but still remains active on the intellectual planes, while seven are still under direct Karmic law. These last act on the man-bearing globes of our chain.[37]

An interpretation of the last quotation is now suggested. The "Great Age" stands for the Manvantara of the Earth planetary system—its evolutionary period of activity. The twelve classes of "Creators" are the Pitris; the first four classes have attained liberation and are therefore even more lofty than the Vairājas who "have reached the Seventh Portal, and have refused Nirvāna."[38] The remaining seven of the twelve classes are represented by the Pitris who are "divided into seven classes . . . three of these are *arūpa,* formless, while four are corporeal."[39]

The Secret Doctrine treats at length on the subject of the "refusal to create," for as already observed, this is a theme which appears again and again in Hindu scriptures. Therefore, another aspect of the subject as to why the Solar Pitris "would not create" is now presented. It deals with the statement that the Agnishvātta Pitris are devoid of the creative fire, whereas the Barhishads, or Lunar Pitris, who created the forms for humanity possessed the creative fire.

> Esoteric philosophy explains the original qualifications as being due to the difference between the natures of the two classes: the Agnishvātta Pitris are devoid of fire (i.e., of creative passion), because too divine and pure, whereas the Barhishad, being the lunar spirits more closely connected with Earth, became the creative Elohim of form, or the Adam of dust.[40]

---

37. Ibid. II:77; III:87, 6-vol. ed.; II:81, 3rd ed.

38. Ibid. V:543, 6-vol. ed.

39. Ibid. II:91; III:100, 6-vol. ed.; II:96, 3rd ed.

40. Ibid. II:78; III:87, 6-vol. ed.; II:81, 3rd ed.

That is to say, a physical form suitable for existence on Earth; the Elohim here being equivalent to the Lunar Pitris.

The significant point is this: whereas the Lords of the Flame were regarded as being devoid of the creative fire, nevertheless

> . . . it was those who would not multiply, who sacrificed themselves to the good and salvation of *Spiritual Humanity.* For, to complete the *septenary man,* to add to his three lower principles and cement them with the spiritual Monad—which could never dwell in such a form otherwise than in an *absolutely latent state*—two connecting principles are needed: *Manas* and *Kāma.*[41]

Manas signifies the mind-principle; Kāma, the energic principle. The two principles thus constitute the personality. The three lower principles are: Prāna, the vital principle; Linga-śarīra, the model vehicle; Sthūla-śarīra, the physical vehicle.

The difference between the assistance provided by the two classes of Pitris—the Corporeal and the Incorporeal—is clearly indicated in the following passage: "the one who builds his external form" designates the Lunar Pitris; "the other" class, the Solar Pitris.

> Each class of Creators endows man with what it has to give: the one builds his external form; the other gives him its essence, which later on becomes the Human *Higher Self* owing to the *personal exertion of the individual;* but they could not make men as they were themselves—perfect, because sinless; sinless, because having only the first, pale shadowy outlines of attributes, and these all perfect—from the human standpoint—white, pure and cold as the virgin snow. Where there is no struggle, there is no merit. Humanity, "of the Earth earthy," was not destined to be created by the angels of the first divine Breath: therefore they are said to *have refused* to do so, and man had to be formed by more material creators, who, in their turn, could give only what they had in their own natures, and no more.[42]

---

41. Ibid. II:79; III:88, 6-vol. ed.; II:83, 3rd ed.

42. Ibid. II:95; III:104, 6-vol. ed.; II:99-100, 3rd ed.

The "more material creators" are, of course, the Lunar Pitris; whereas the Lords of the Flame are referred to as "the angels of the first divine Breath."

In *The Secret Doctrine* the Lords of the Flame are definitely associated with the Agnishvātta Pitris in these words:

> These are then those "Flames" (the *Agnishvātta*) who, as shown in Śloka 13, "remain behind" instead of going along with the others to create men on Earth. But the true meaning is that most of them were destined to incarnate as the Egos of the forthcoming crop of Mankind.[43]

Whence Came the Lords of the Flame?

A question which has been raised concerning the coming of the Lords of the Flame to the Earth from Venus. In all probability this assumption was formulated because of a statement made in a Commentary from the Book of Dzyan that each one of the Seven Root-Races comes under the regency of one of the Seven Sacred Planets. The Dhyāni of Venus is the Regent of the Third Root-Race, and it was during the fifth subrace of the Third Root-Race that the enlightenment of humanity was brought about by the Lords of the Flame. Here are the words of the Commentary:

> Like each of the seven regions of the Earth, each of the seven First-born (the primordial human groups) receives its light and life from its own especial Dhyāni—spiritually, and from the palace (house, the planet) of that Dhyāni physically; so with the seven great Races to be born on it. The first is born under the Sun; the second under Brihaspati (Jupiter); the third under Lohitanga (the "fiery-bodied," Venus, or Śukra) . . . .[44]

The Lords of the Flame are also called the Gods of Will, and they are described as the celestial beings who descended to Earth in order to enlighten the human race. A commentary from the Book of Dzyan states:

---

43. Ibid. II:79; III:89, 6-vol. ed.; II:83, 3rd ed.

44. Ibid. II:29; III:41-2, 6-vol. ed.; II:32, 3rd ed.

> Man needs four flames and three fires to become one on Earth. . . . It is those who have deserted the Superior Spheres, the Gods of Will, who complete the Manu of illusion.[45]

The four flames signify the four lower principles (described above) ; the three fires are the immortal principles. Ātman, Buddhi, Manas; whereas the "manu of illusion" means man before he became a self-conscious being. "The Superior Spheres," which were deserted by the Lords of the Flame (the Gods of Will), represent the superior globes of the Earth planetary system.

Śloka 24 of Stanza VII contains the following reference to the Sons of Wisdom—another of the descriptive names of the Lords of the Flame:

> The Sons of Wisdom, the Sons of Night (issued from the body of Brahmā when it became Night), ready for re-birth, came down.[46]

The Sons of Wisdom are called the Sons of Night because they had attained "Wisdom" in a former Manvantara, inasmuch as the Day of Brahmā represents the present Manvantara. They are said to have come down into the present sphere because of having descended from a superior sphere of the Earth planetary system.

The same idea is presented in the Brahmanical teachings in connection with the Lokas, as shown in Diagram V of *The Secret Doctrine*.[47] There the Rūpa Devas (the Gods of Form) are allocated to Maharloka, whereas our human kingdom is portrayed as undergoing its evolution on Bhūrloka, three lokas lower than Maharloka. The Kumāras are situated on Janarloka, and on a still higher loka—Taparloka—are the Vairājas. In *The Secret Doctrine* the Lokas and Talas are represented as states of consciousness. The Lunar Pitris are here termed Rūpa Devas inasmuch as they provided the forms (rūpas) for the monads who were awaiting the forma-

---

45. Ibid. II:57; III:67-8, 6-vol. ed.; II:60, 3rd ed.

46. Ibid. II:161, III:168, 6-vol. ed.; II:170, 3rd ed.

47. The diagram is placed facing page 543 of Volume V.

tion of fit rūpas for the peopling of the Earth. Elsewhere the Lunar Pitris (termed Progenitors) are described as stationed on Maharloka, quoting a Commentary:

> Having projected their shadows and made men of one element (ether), the progenitors re-ascend to Maharloka, whence they descend periodically, when the world is renewed, to give birth to new men.[48]

There is yet another passage which graphically describes the activities of both the Solar and Lunar Pitris. The Solar Pitris are therein termed "Celestial Ancestors" whereas the Lunar Ancestors are merely designated "Pitris" (Fathers). Both Pitris are described not as coming from another planet, but simply as "stepping in on this our plane," signifying that they came from higher planes on which the superior globes of the Earth planetary system are situated:

> Therefore, when the hour strikes . . . for the formation of the Perfect Man—rudimentary man of the first 2 1/2 Races being only the *first,* gradually evolving into *the most perfect of mammals*—the Celestial "Ancestors" (Entities from preceding worlds, called in India the Śishtas) step in on this our plane, as the Pitris had stepped in before them for the formation of the physical or animal-man, and incarnate in the latter. Thus the two processes—for the two *creations,* the animal and the divine man—differ greatly. The Pitris shoot out from their ethereal bodies, still more ethereal and shadowy similitudes of themselves, or what we should now call "doubles" or "astral forms," in their own likeness. This furnishes the Monad with its first dwelling, and blind matter with a model around which to build henceforth.[49]

48. SD II:92; III:101, 6-vol. ed.; II:96, 3rd ed.
49. Ibid. I:248; I:293, 6-vol. ed.; I:268, 3rd ed.

14.

THE SEVEN GROUPS OF MONADS

THE FOURTEENTH śloka of the Stanzas of Dzyan specifies that there are seven hosts of Lunar Pitris. Therefore it is logical to assume that seven groups of monads are involved in the peopling of the Earth. In fact the account is emphatic that this is what occurred. However, it should be borne in mind that the seven groups associated with the peopling of the Earth are distinct from the "three Classes of Monads" which have been referred to in this narrative. Nor should these seven groups of monads be equated to the Seven Root-Races. Briefly described, the three classes of monads are:

1. The most developed monads: those who led the human kingdom during the three preceding Rounds.

2. Those monads who attain the human stage during the three and a half Rounds and form the vast majority of the present humanity.

3. Those monads who will not reach the human stage at all during this cycle or Round.

The significant point in connection with the *first* class of monads is that although they had completed the seven-round cycle of evolution on the Lunar Chain and had graduated from the human kingdom on the Moon, they were still so strongly connected with the monads of the second class—now the present human kingdom on Earth—that they were called upon to provide the rūpas to be occupied by the second class of monads when they were due to incarnate on this Earth during the present Fourth Round.

This *second* class of monads is now pursuing the evolutionary cycle in the human kingdom on this Earth. The monads are themselves of two classes: (a) those who were in the evolutionary cycle in the human kingdom on the Lunar Chain who did not graduate; and (b) those who were ready to enter the human stage on the Earth Chain at any time during the preceding three and a half Rounds.

The *third* class of monads are those who were not ready to enter the human kingdom at the midpoint of the Fourth Round on Earth.

The Activities of the Lunar Pitris

Sloka 13 recounts how the Lunar Pitris responded to the commands of the great Chohans:

> They (the Moon-gods) went, each on his allotted land: seven of them, each on his lot.[1]

Thus there were seven hosts of Pitris, each one of the seven hosts proceeding to a particular zone on Earth. A Commentary from the Book of Dzyan offers this explanation:

> Like each of the seven regions of the Earth, each of the seven First-born (the primordial human groups) receives its light and life from its own especial Dhyāni—spiritually, and from the palace (house, the planet) of that Dhyāni physically.[2]

The "seven First-born," described as the "Primordial human groups," are indeed the monads, the seven groups of monads, who are responsible for peopling the Earth. Each one of the seven groups is under the regency of "its own especial Dhyāni." Just as are the seven regions of the Earth, the seven Root-Races, the seven principles of man, and the seven globes of the Earth Chain, so there are the seven zones, each to receive its own host of Pitris.

Attention is particularly directed to the fact that the specified seven primordial human groups are not the same as the seven Root-Races, and the two *must not be confused.* The seven primordial human groups represent seven distinct

---

1. SD, II:77; III:86, 6-vol. ed.; II:81, 3rd ed.

2. Ibid. II:29; III:41, 6-vol. ed.; II:32, 3rd ed.

assemblages of monads. The seven Root Races, on the other hand, are the seven major developmental stages of evolution accomplished in turn on each one of the seven globes of the terrene planetary system. Each of the seven primordial human groups, or seven groups of monads, is obliged to experience each of the seven root race stages of evolution in serial order, on each one of the seven globes, so as to accomplish the complete seven circuits or Rounds of the evolutionary pilgrimage, or the Circle of Necessity.

Diagram Representing the Seven Primordial Groups

A diagram may be used to illustrate the idea presented in the Commentary. The large inclusive circle represents the Earth (Globe D). In this circle are seven smaller circles, each indicating one of the seven zones or "lots" to which one of the hosts of the Lunar Pitris went. As each zone is under the regency of a Watcher, or Dhyāni, the appropriate astrological symbol for each presiding Dhyāni is placed next to each one of the seven small circles. The symbols stand for the seven Sacred Planets: (1) the planet for which the Sun is the substitute: (2) Jupiter; (3) Venus; (4) Saturn; (5) Mercury; (6) Mars; and (7) the planet for which the Moon is the substitute.

Above the large circle is a smaller circle which stands for the Moon. Seven rays are shown extending from the sphere of the Moon, representing the seven hosts of the Lunar Pitris who accomplished the Circle of Necessity on the Lunar Chain and graduated from the lunar human kingdom. Just as each one of the seven human groups is under the regency of a Dhyāni, so is each host of Pitris.

Śloka No. 13 describes how each one of the seven hosts of Pitris went to his allotted lot and produced the rūpas for the human monads. Śloka No. 14 continues the account under the name of the Will-born Lords:

> The Seven Hosts, the "Will-born Lords," propelled by the Spirit of Life-giving, separate men from themselves, each on his own zone.[3]

3. Ibid. II:17; III:29, 6-vol. ed.; II:18, 3rd ed.

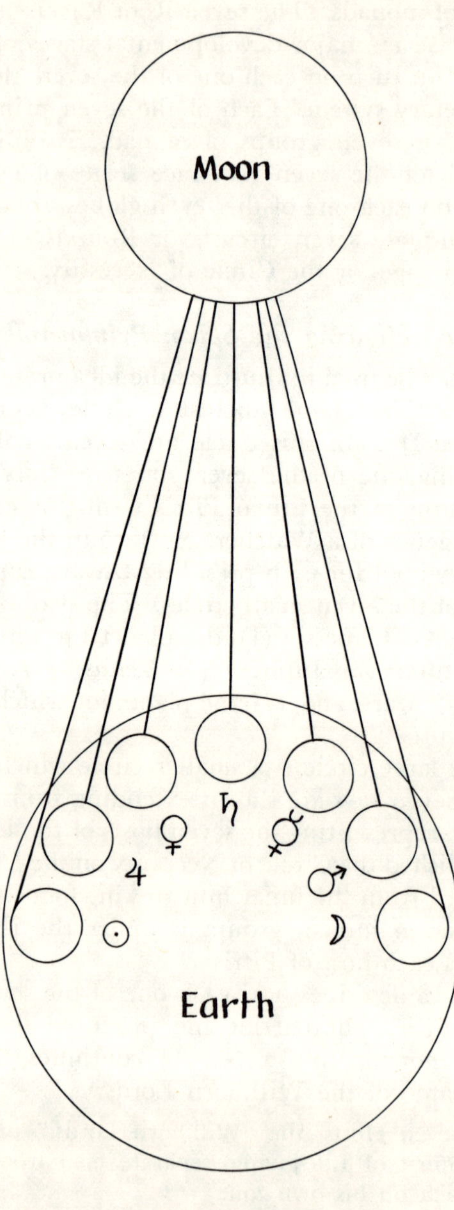

DIAGRAM 10

The Secret Doctrine gives an explanation of the śloka, adding that the "Spirit of Life-giving" is Fohat:

> Occult philosophy . . . teaches that the first human stock was projected by higher and semi-divine Beings out of their own essences. If the latter process is to be considered as abnormal or even inconceivable—because obsolete in Nature at this point of evolution—it is yet proven possible on the authority of certain "Spiritualistic" FACTS.[4]
>
> For who of those who have witnessed the phenomenon of a materializing form oozing out of the pores of a medium or, at other times, out of his *left side,* can fail to credit the possibility, at least, of such a *birth*?[5]

It should be borne in mind that the rūpas (or forms) which were projected by the Lunar Pitris (described above as "semi-divine Beings"), were not at all similar to the physical bodies of present-day humanity. For the quotation continues:

> . . . man was not "created" the complete being he is now, however imperfect he still remains. There was a spiritual, a psychic, an intellectual, and an animal evolution, from the highest to the lowest, as well as a physical development—from the simple and homogeneous, up to the more complex and heterogeneous; though not quite on the lines traced for us by the modern evolutionists. This double evolution in two contrary directions, required various ages, of divers natures and degrees of spirituality and intellectuality, to fabricate the being now known as man. Furthermore, the one absolute, ever acting and never erring law, which proceeds on the same lines from one eternity (or Manvantara) to the other—ever furnishing an ascending scale for the manifested, or that which we call the great illusion (Mahā-Māyā), but plunging Spirit deeper and deeper into materiality on the one hand, and then *redeeming it through flesh* and liberating it—this law,

---

4. Ibid. II:87; III:96, 6-vol. ed.; II:91, 3rd ed.

5. Ibid. II:86; III:95-6, 6-vol. ed.; II:91, 3rd ed.

> we say, uses for these purposes the Beings from other and higher planes, men, or Minds (Manus), in accordance with their Karmic exigencies.[6]

The "one absolute ever active and never erring law" signifies the obligatory pilgrimage which must be undertaken by the monad through the Circle of Necessity, as outlined in the third fundamental proposition in *The Secret Doctrine.*

The creative activity of the Lunar Pitris is continued in śloka No. 15 of Stanza IV, the first sentence of which reads:

> Seven times seven Shadows of future men were born, each of his own color and kind.[7]

The Chhāyā Race

When śloka No. 15 was explained, three words were added to the first sentence. After the word "shadows" the Sanskrit equivalent word "chhāyās" was placed. The word is here applied to the first humans who peopled the Earth, and they are regarded as the First Root-Race, which is also termed the Chhāyā Race, representing the first stage of evolutionary development on this Earth.

The second word, *amānasa,* was inserted after "future men." The Sanskrit word *mānasa* is the adjectival form of *manas,* meaning "mental." The negative particle *"a,"* meaning "not," establishes the fact that the first humans were "non-mental," because the rūpa which the Lunar Pitris "created" was not provided with a functioning mental principle, *Manas.*

The third word "complexion" was added after the word "color." Thus each zone had its specific complexion as well as its "own kind."

An explanation is provided in regard to the "shadows" emanated by the Lunar Pitris:

> They threw off their "shadows" or *astral bodies*—if such an ethereal being as a "lunar Spirit" may be sup-

---

6. Ibid. II:87-8; III:96-7, 6-vol. ed.; II:92, 3rd ed.

7. Ibid. II:17; III:29, 6-vol. ed.; II:18, 3rd ed.

> posed to rejoice in an astral, besides a hardly tangible body. In another Commentary it is said that the "Ancestors" *breathed* out the first man, as Brahmā is explained to have breathed out the *Suras* (Gods), when they became "Asuras" (from *Asu,* breath). In a third it is said that they, the newly-created men, "were the shadows of the Shadows."[8]

An interesting point is brought forward in connection with the reference which is made to the "Suras when they became Asuras" and the words "from *Asu* breath" were added. In the *Rig-Veda* Asuras are regarded as divine beings, whereas in later Hindu scriptures the Asuras are termed "demons," thereby rendering the Sanskrit word as compounded of the negative particle "a"—not; and "sura"—god; hence "not a god—or demon." So that we find:

> Every ancient theogony without exception—from the Aryan and the Egyptian down to that of Hesiod—places, in the order of Cosmogonical evolution, Night before the Day; even *Genesis,* where "darkness is upon the face of the deep" before "the *first day.*" The reason for this is that every Cosmogony—except in the Secret Doctrine—begins by the "Secondary Creation" so-called, to wit, the *manifested* Universe, the Genesis of which has to open by a marked differentiation between the eternal Light of *Primary* Creation, whose mystery must remain for ever "Darkness" to the prying finite conception and intellect of the profane, and the Secondary Evolution of manifested visible nature.[9]

The same idea is applicable to the Solar and the Lunar Pitris. Both of these divine "Ancestors" are to be regarded as having accomplished their evolutionary status in previous planetary systems and not on this Earth planetary system.

It has been suggested by some writers that "ethereal body" would be more suitable than "astral body" for the Sanskrit term *Linga-śarīra,* "since astral body is also used to translate the term Kāma-rūpa"—literally "the model" or pattern, *linga,*

---

8. Ibid. II:86; III:95, 6-vol. ed.; II:90, 3rd ed.

9. Ibid. II:59; III:69-70, 6-vol. ed.; II:62, 3rd ed.

which "dissolves away," *śarīra* (derived from the verbal root, *srī,* to dissolve).

Further explanation of the "astral bodies" is provided in this passage:

> The Pitris shoot out from their ethereal bodies, still more ethereal and shadowy similitudes of themselves, or what we should now call "doubles," or "astral forms," in their own likeness. This furnished the Monad with its first dwelling, and blind matter with a model around and upon which to build henceforth. But *Man is still incomplete.*[10]

Man Is Still Incomplete

In spite of the ethereal vehicle which was produced by the Lunar Pitris, man is still regarded as incomplete. The significance of this is further considered in the next chapter. Yet a form without a monad is indeed naught but a shadow, a *bhūta,* as described below in the continuation of śloka 15 of the fourth Stanza of Dzyan:

> Each inferior to his father. The fathers, the boneless, could give no life to beings with bones. Their progeny were Bhūta, with neither form nor mind. Therefore they are called the Chhāyā.[11]

The word "each" refers to "every shadow" which had been produced by each Pitri. The "shadows" are referred to as being inferior to the fathers because the Pitris had concluded the Circle of Necessity, or the seven-round cycle of evolution on the Lunar Chain; whereas the shadows represent only the *vehicles* which the monads of the human kingdom are to assume, so that they may continue their evolutionary progress of the fourth circuit—or Fourth Round—on the fourth sphere or globe of the sevenfold Earth planetary system.

The Pitris were not able to produce rūpas with bones because their own rūpas (forms) were without bones. Con-

---

10. Ibid. I 248; I:293, 6-vol. ed.; I:268, 3rd ed.

11. Ibid. II:17; III:29, 6-vol. ed.; II:18, 3rd ed.

sequently their progeny are described as being "bhūta"—a word signifying a phantom;[12] that is to say, without a physical form although the rūpa would have an ethereal form, and the mind would not be functioning. To clarify this statement a Commentary from the Book of Dzyan is added:

> It is from the material Worlds that descend they, who fashion physical man at the new Manvantaras. They are inferior Lha (Spirits), possessed of a dual body (an astral within an ethereal form). They are the fashioners and creators of our body of illusion.[13]

The rūpa provided by the Lunar Pitris is termed a "body of illusion" because (1) it is subject to continual change, indicated by the words Nitya-Sarga and Nitya-Pralaya. *Nitya* is a Sanskrit word meaning continual, perpetual, eternal; *sarga,* from the verbal root *srij,* signifies emission or creation of matter; *pralaya,* dissolution. These terms apply to the processes of building up and breaking down, which go on incessantly in an aggregation of living cells and substance, such as the physical body. (2) The rūpa changes with every Root Race, regarded as a major developmental stage of evolution.

The "new Manvantaras" *in this case* signify the new periods of activity which occur during the Rounds, when the monads pass from one globe to the next of the series of seven globes of the planetary system. Before the start of each "new Manvantara" there is a waiting period, which is a period of rest.

---

12. A significant footnote in *The Secret Doctrine* is worthy of note: It is not clear why "*Bhūtas*" should be rendered by the Orientalists as meaning "evil Spirits" in the Purānas. In the *Vishnu-Purāna,* Book I, ch. 5, the Śloka simply says: "Bhūtas—fiends, frightful from being monkey-colored and carnivorous"; and the word in India now means *ghosts,* ethereal or *astral* phantoms, while in esoteric teaching it means *elementary* substances, something made of attenuated, non-compound essence, and, specifically, the astral *double* of any man or animal. In this case these primitive men are the *doubles* of the first ethereal Dhyānis or Pitris. (II:102; III:110-1, 6-vol. ed.; II:107, 3rd ed.)

13. SD II:57; III:67, 6-vol. ed.; II:60, 3rd ed.

Notice the name by which those who first peopled the Earth are known: the Chhāyā Race. A Commentary from the Book of Dzyan explains:

> The Dhyānis (Pitris) are those who have evolved their BHŪTA (doubles) from themselves, which RŪPA (form) has become the vehicle of monads (seventh and sixth principles) that had completed their cycle of transmigration in the three preceding Kalpas (Rounds). Then, they (the astral doubles) became the men of the first Human Race of the Round. But they were not complete, and were senseless.[14]

Again mention is made of the incompleteness of man. The seventh and sixth principles are Ātman, the divine spirit, and Buddhi, the discriminating principle. The technical meaning of the term "monad" signifies the union of Ātma-Buddhi. The cycle of transmigration is the complete circuit of all the seven globes of the planetary system, involving seven major developmental stages of evolution on each one of the seven globes. One circuit of the seven globes is known as a Round, and the time-period of a Round is one Kalpa. The Round referred to here is the Fourth Round, the fourth Kalpa, which follows three preceding Kalpas.

The Commentary continues its description of man in the preceding Kalpas and also in the present Fourth Round on this Earth:

> The inner, now concealed, man, was then (in the beginnings) the external man. The progeny of the Dhyānis (Pitris), he was "the son like unto his father." Like the lotus, whose external shape assumes gradually the form of the model within itself, so did the form of man in the beginning evolve from within without. After the cycle in which man began to procreate his species after the fashion of the present animal kingdom, it became the reverse. The human foetus follows now in its transformations all the forms that the physical frame of man had assumed throughout the three Kalpas (Rounds) during the tentative efforts at plastic formation around

14. Ibid. I:183; I:235, 6-vol. ed.; I:206, 3rd ed.

> the monad by senseless, because imperfect, matter, in her blind wanderings. In the present age, the physical embryo is a plant, a reptile, an animal, before it finally becomes man, evolving within himself his own ethereal counterpart, in his turn. In the beginning it was that counterpart (astral man) which, being senseless, got entangled in the meshes of matter.[15]

"The inner, now concealed, man" signifies the Linga-śarīra, the model body. The reference to the lotus is explained by a passage in *The Secret Doctrine*:

> . . . the Lotus-seed contains within itself a perfect miniature of the future plant, which typifies the fact that the spiritual prototypes of all things exist in the immaterial world before those things become materialized on Earth.[16]

The next passage from *The Secret Doctrine* admirably clarifies the concept of prototypes, and the application of the principle to the form of man:

> Everything that *is, was,* and *will be,* eternally *IS,* even the countless forms, which are finite and perishable only in their objective, not in their ideal Form. They existed as Ideas, in the Eternity, and, when they pass away, will exist as reflections. Neither the form of man, nor that of any animal, plant or stone has ever been *created,* and it is only on this plane of ours that it commenced "becoming," i.e., objectivizing into its present materiality, or expanding *from within outwards,* from the most sublimated and supersensuous essence into its grossest appearance. Therefore *our* human forms have existed in the Eternity as astral or ethereal prototypes; according to which models, the Spiritual Beings (or Gods) whose duty it was to bring them into objective being and terrestrial Life, evolved the protoplasmic forms of the future *Egos* from *their own essence.* After which when this human *Upādhi,* or basic mold was ready, the natural terrestrial Forces began to work on

---

15. Ibid. I:184; I:235, 6-vol. ed.; I:206, 3rd ed.

16. Ibid. I:58; I:127-8, 6-vol. ed.; I:88, 3rd ed.

> those supersensuous molds which contained, besides their own, the elements of all the past vegetable and future animal forms of this globe in them.
>
> Occultism teaches that no form can be given to anything, either by nature or by man, whose ideal type does not already exist on the subjective plane. More than this; that no such form or shape can possibly enter man's consciousness, or evolve in his imagination, which does not exist in prototype, at least as an approximation.[17]

This would seem to indicate that there are other and better forms, the prototypes of which are in existence although not yet manifest. At the appropriate time and after evolutionary progress has been made, man will then assume better forms, in which he can express his spiritual faculties and powers, just as our present human form is the result of changes which have taken place ever since the time of the Chhāyā Race up to our present Fifth Race. A passage on this theme is now presented:

> . . . man—or rather his Monad—has existed on the earth from the very beginning of this Round. But, up to our own Fifth Race, the external shapes which covered those divine astral doubles changed and consolidated with every sub-race; the form and physical structure of the fauna changing at the same time, as they had to be adapted to the ever-changing conditions of life on this globe during the geological periods of its formative cycle. And thus shall they go on changing with every Root Race and *every chief sub-race* down to the last one of the Seventh [Root Race] in this Round.[18]

As there are seven Root-Races, each one of which undergoes a major evolutionary stage of development during its specific time-period, it should be clarified that there are seven subraces for each Root-Race. Each subrace accomplishes a minor evolutionary stage of development during the time-period of one Root-Race.

---

17. Ibid. I:282; I:324-5, 6-vol. ed.; I:303, 3rd ed.

18. Ibid. I:183-4; I:235, 6-vol. ed.; I:206, 3rd ed.

15.

HOW MAN WAS COMPLETED

THUS far the Stanzas of Dzyan have shown how the vehicle of man was produced, and that it was regarded as a bhūta "with neither form nor mind." Because of this, the first monads who peopled the Earth were called the Chhāyā, the Shadow Race, because their rūpas (forms) were not physical. Śloka 16 of Stanza IV continues the theme of man's incompleteness with two questions: "How are the Manushya born?" and "The Manus with minds, how are they made?" Manushya is the Sanskrit word for a human being whereas Manu here is a thinking being, and this is emphasized by the words "Manus with minds."

The process of completion of man occurred long after he actually came to the Earth, and so its full treatment is beyond the scope of this work. However, it is here referred to briefly because a commentary from the Book of Dzyan answers the two questions from śloka 16 and links man's completion with the subject of his septenary constitution. *The Secret Doctrine* introduces the Commentary with these words:

> This Stanza we shall see is very suggestive. It explains the mystery of, and fills the gap between, the informing principle in man—the HIGHER SELF or human Monad—and the animal Monad, both one and the same, although the former is endowed with *divine* intelligence, the latter with *instinctual* faculty alone.[1]

1. SD, II:102-3; III:111, 6-vol. ed.; II:108, 3rd ed.

In *The Key to Theosophy* H. P. Blavatsky was more explicit in her use of terms concerning the seven principles of man. She made a distinction between the Higher Self and the *human monad.* The Higher Self was there termed Ātma-Buddhi (p. 121), whereas the human monad or the Reincarnating Ego was Buddhi-Manas. Then "the informing principle in man" is often called the individuality with its divine intelligence, whereas the animal monad, Kāma-Manas, is called the personality, or again the "human animal" endowed "with instinctual faculty alone." Of especial significance is the statement that the human monad and the animal monad are "both one and the same." In other words, the individuality and the personality are not separate discrete entities, but "one and the same." When death takes place, the good, the true and the beautiful (to use the Platonic expression) are indrawn into the Reincarnating Ego during the "gestation period"[2] culminating in the "second death." This indrawing process with its accumulated results produces the new personality at the next rebirth of the Reincarnating Ego. The difference between the individuality and the personality is raised in this question, which is answered by quoting a Commentary from the Book of Dzyan:

> How is the difference to be explained, and the presence of that HIGHER SELF in man accounted for?[3]

The Sons of Mahat

> "The Sons of MAHAT are the quickeners of the human Plant. They are the Waters falling upon the arid soil of latent life, and the Spark that vivifies the human animal. They are the Lords of Spiritual Life eternal. . . .
> "In the beginning (in the Second Race) some (of the Lords) only breathed of their essence into Manushya (men); and some took in man their abode."[4]

---

2. The gestation state is described in *The Mahatma Letters to A. P. Sinnett,* pp. 103-5, 2nd ed.; pp. 101-3, 3rd ed.

3. SD II:103; III:111, 6-vol. ed.; II:108, 3rd ed.

4. Ibid.

"Sons of Mahat" is another name for the Mānasaputras, the Solar Pitris: *putras* signify sons; *Mahat,* literally "the Great" mind-principle, the source of Manas. Man prior to the awakening of the mind principle is referred to as the human animal—a Manushya; and without the functioning of Manas likened to arid soil without water from the fountain of Wisdom. Man becomes a Manu, a thinking being, when vivified by the spark of Manas. Hence the Sons of Mahat are referred to as the Lords of Spiritual Life, because they enabled man to attain spirituality instead of functioning only instinctually. Then, too, because of "taking their abode in man" they speeded up man's evolution, although the full awakening did not occur until later eras, during the Third Race and not in the Second Race. As *The Secret Doctrine* explains it:

> This shows that not all men became incarnations of the "divine *Rebels,*"[5] but only a few among them. The remainder had their fifth principle [Manas] simply quickened by the spark thrown into it, which accounts for the great difference between the intellectual capacities of men and races. Had not the "sons of Mahat," speaking allegorically, skipped the intermediate worlds, in their impulse toward intellectual freedom, the animal man would never have been able to reach upward from this earth, and attain through self-exertion his ultimate goal. The cyclic pilgrimage would have to be performed through all the planes of existence half unconsciously, if not entirely so, as in the case of the animals. It is owing to this rebellion of intellectual life against the morbid inactivity of pure spirit, that we are what we

---

5. Why some of the Sons of Mahat are called Divine Rebels is best explained by quoting another passage:

> The Secret Doctrine teaches that the Fire-Devas, the Rudras, and the Kumāras, the "Virgin-Angels," . . . the Divine "Rebels," . . . preferred the *curse of incarnation* and the long cycles of terrestrial existence and rebirths, to seeing the misery (even if *unconscious*) of the beings (evolved as shadows out of their Brethren) . . . This voluntary sacrifice of the Fiery Angels, whose nature was *Knowledge* and *Love,* was construed by the exoteric theologies into a statement that shows "the rebel angels hurled down from heaven into the darkness of Hell"—our Earth. (*The Secret Doctrine* II:246; III:248-9, 6-vol. ed.; II:256-7, 3rd ed.)

are—self-conscious, thinking men, with the capabilities and attributes of Gods in us, for good as much as for evil. Hence the REBELS are our saviors. Let the philosopher ponder well over this, and more than one mystery will become clear to him. It is only by the attractive force of the contrasts that the two opposites—Spirit and Matter—can be cemented on Earth, and, smelted in the fire of self-conscious experience and suffering, find themselves wedded in Eternity.[6]

The Seven Principles of Man

Primary consideration has thus far been given to the second series of the Stanzas of Dzyan narrating the account of the peopling of the Earth. In order to deal with the subject of the "completion of man" reference is now made to the first series of the Stanzas—to Stanza VII, śloka 3—where mankind is termed symbolically a "man-plant," *Saptaparna.* The Sanskrit word *Saptaparna* literally means "the seven-leaved;" *sapta,* seven; *parna,* leaf. Then, too, the three highest principles of man are included therein:

> When the one becomes two, the threefold appears, and the three are one; and it is our thread, O Lanoo, the heart of the man-plant called Saptapárna. (Stanza VII, śloka 3)[7]

Making use of the anthropological meaning rather than the cosmical one, which is also applicable, the monadic essence, Ātman—the divine spirit, the one—becomes two when Buddhi is emanated from Ātman. Buddhi is the intuition, also regarded as the discriminating principle. These two principles constitute the monad. "The threefold appears" when the monad emanates Manas, the mind-principle. When these three immortal principles are linked into or regarded as one, they represent the heart of the saptaparna, the seven-leaved "man-plant." The thread stands for the enduring portion of man's seven principles, for it continues

6. SD II:103; III:111-12, 6-vol. ed.; II:108, 3rd ed.

7. Ibid. I:34; I:98, 6-vol. ed.; I:65, 3rd ed.

from life to life on Earth. In the Vedāntic philosophy the very same idea is expressed in the word Sūtrātman, literally the thread-self, the golden cord which enables man to manifest on earth, and which links the immortal Self with the impermanent personal vehicles.

The fourth śloka continues the reference to the threefold thread:

> It is the root that never dies; the three-tongued flame of the four wicks. The wicks are the sparks, that draw from the three-tongued flame shot out by the seven—their flame—the beams and sparks of one moon reflected in the running waves of all the rivers of earth. (Stanza VII of the first series)[8]

The Secret Doctrine explains:

> The "Three-tongued flame" that never dies is the immortal spiritual triad—the Ātma-Buddhi and Manas—the fruition of the latter assimilated by the first two after every terrestrial life. The "four wicks" that go out and are extinguished, are the four lower principles, including the body.[9]

The very same idea is alluded to in another ancient scripture: in the Egyptian Book of the Dead:

> "I am the three-wicked Flame and my wicks are immortal," says the defunct. "I enter into the domain of Sekhem [the God whose arm sows the seed of action produced by the disembodied soul] and I enter the region of the Flames who have destroyed their adversaries," i.e., got rid of the sin-creating "four wicks."[10]

The "sin-creating four wicks" are the mortal four lower principles of the sevenfold constitution. However, the declaration by the deceased that his "wicks are immortal" alludes to the life-atoms which compose the rūpa and these are indeed immortal and pursue their peregrinations in their appropriate kingdoms during the after-death state of man.

---

8. Ibid. I:34; I:98, 6-vol. ed.; I:65, 3rd ed.

9. Ibid. I:237; I:282, 6-vol. ed.; I:257, 3rd ed.

10. Ibid., quoting the Book of the Dead, ch. i, 7.

Those atoms which move by kinetic energy are termed "life-atoms," the aggregations of which compose the vehicles of the kingdoms of nature, that is, the mineral, plant, animal and human kingdoms. These life-atoms are constantly in motion, engaged in the process of upbuilding and sustaining the vehicles of all the kingdoms. In the case of the human kingdom, when death occurs the life-atoms composing the physical body are released and consequently they travel to their appropriate kingdoms, from which they were drawn in order to form the physical body.

Śloka 4 (of Stanza VII of the first series) is of particular interest because, when commenting upon it, an exquisite passage was written by H. P. Blavatsky, showing her versatility and talent in descriptive writing:

> Just as milliards of bright sparks dance on the waters of an ocean above which one and the same moon is shining, so our evanescent personalities—the illusive envelopes of the immortal MONAD-EGO—twinkle and dance on the waves of Māyā. They last and appear, as the thousands of sparks produced by the moon-beams, only so long as the Queen of the Night radiates her lustre on the running waters of life; the period of a Manvantara; and then they disappear, the beams—symbols of our eternal Spiritual Egos—alone surviving, remerged in, and being, as they were before, one with the Mother-Source.[11]

Here, profound philosophy has been expressed beautifully and symbolically. In the following philosophical exposition, the fourth śloka is also included.

The root that never dies signifies Ātman—the divine counterpart, or Spirit, in the sevenfold constitution of a human being (commonly referred to as the seven principles of man). It is represented as a three-tongued flame, since Ātman is unable to manifest in physical realms without a vehicle (to be further explained under the term "Monad-Ego"). The three-tongued flame, then, stands for man's immortal triad: Ātma-Buddhi-Manas. The four wicks sym-

---

11. Ibid., I:237; I:283, 6-vol. ed.; I:258, 3rd ed.

bolize the lower quaternary, the perishable components of the personality: (1) Sthūla-śarīra, the physical body, or vehicle; (2) Linga-śarīra, the model body; (3) Prāna, the life principle or vitality; (4) Kāma, the desire principle, providing the energic stimulus in man. The wicks are likened to sparks in śloka 4 because during an earth-life a spark (or the evanescent personality) is enabled to exist so long as the wick is sustained by means of its contact with the three-tongued flame. In its turn the wick transmits the good, the true, and the beautiful to Higher Manas (Manas-Taijasī) when the four wicks are extinguished at death. It is Lower Manas—the mind principle, in conjunction with the desire principle, Kāma—along with the two other wicks which constitute the evanescent personality during a life-time on Earth. This personality is described as an illusive envelope of the immortal "Monad-Ego." Notice the compound word used here for the three-tongued flame, the root that never dies—the Monad (Ātman and Buddhi) plus the Ego, standing for Higher Manas. It is the evanescent personalities, appearing and disappearing, which twinkle and dance on the waves of Māyā. The Queen of the Night is, of course, the moon, and Māyā signifies earth-life, one personality following another in incarnation on Earth.

Assuredly, during the period of a Manvantara—the seven-round period of activity of the planetary system of Earth (Bhūmi)—the coming and going of the illusive envelopes may indeed appear as thousands of moon-beam sparks, yet the *spiritual aroma* of each personality is garnered during the after-death state, and is assimilated into the immortal triad. As Manas-Taijasī, this aroma constitutes that which is experienced in the dream-like Devachanic interlude. This aroma, so to speak, becomes the individual quality of the spiritual Ego. The closing words of the descriptive passage allude to the Solar Pralaya, following a Solar Manvantara, during which the monad re-merges for a period with its Mother-Source, Paramātman, the Source of Ātman.[12]

---

12. The exposition of the passage from *The Secret Doctrine,* as well as śloka 4 of Stanza VII, is taken from the author's book *H. P. Blavatsky, Tibet and Tulku,* pp. 176-7.

The seven principles of man are further symbolically described in one of the Commentaries from the Book of Dzyan, where the seven principles are also divided into the mortal and the immortal components:

> Man needs four flames and three fires to become one on Earth, and he requires the essence of the forty-nine fires to be perfect.[13]

The three fires are the three immortal principles of the sevenfold constitution, the higher triad:

Ātman	Spirit, the Universal principle
Buddhi	the Discriminating principle
Manas	the Mind principle

The four flames are the four mortal principles in the sevenfold constitution of man. These are the lower quaternary:

Kāma	the Desire principle
Prāna	the Life principle
Linga-śarīra	the Model body
Sthūla-śarīra	the Physical body

The Forty-Nine Fires

The complete evolutionary development of the seven principles is the attainment of perfect manhood. Each of the seven principles has seven subprinciples, each in its turn reflecting the characteristics of the main principles. This will entail the unfoldment of the forty-nine Fires, to use the expression of the Commentary. The forty-nine Fires may well be illustrated on a diagram.

To attain the perfection indicated in the words of the Commentary, the essence of each one of the principles must be brought to full development. Such, indeed, is the goal. The seven Rounds provide the opportunity for accomplishing the complete unfoldment of all the forty-nine Fires, whereas one Round enables one principle to be developed

---

13. SD II:57; III:67, 6-vol. ed.; II:60, 3rd ed.

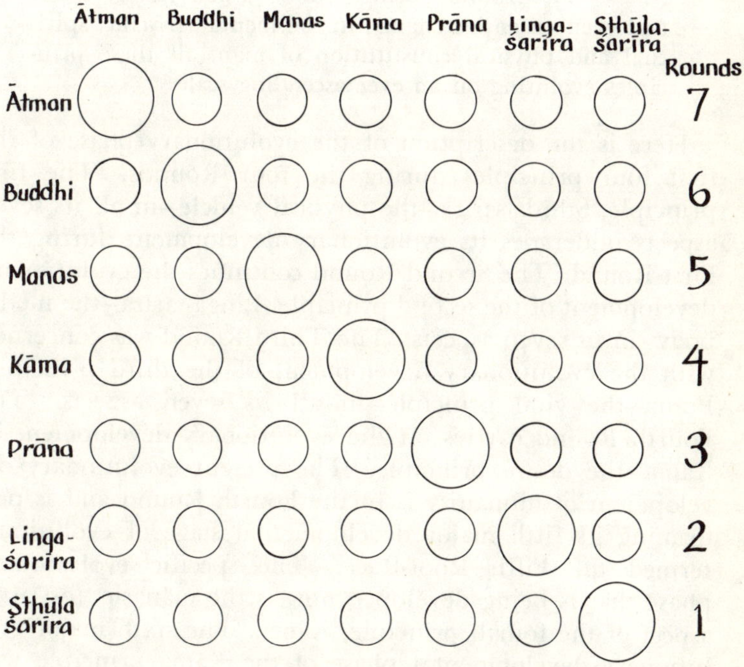

DIAGRAM 11

This diagram is based upon the words of the Commentary, quoted herewith, in regard to the Four Flames and the Three Fires and the essence of the forty-nine fires. This signifies that each one of the seven principles has seven subprinciples, thus forming the forty-nine Fires. Further, that one principle is developed in full evolutionary capacity by means of its seven subprinciples or aspects during one Round. On the left-hand side of the diagram the seven principles are indicated; on the right-hand side are placed the seven Rounds; the first Round is placed on the lowest line. The seven principles are also indicated at the top of the diagram. The small circles on each of the seven horizontal lines signify the subprinciples or aspects of the seven main principles. The large circles represented diagonally indicate the principle that is to be evolutionally developed during that Round. The complete evolutional development of the seven principles is indicated by the seventh Round horizontal line—the ātmic aspect of Ātman—thus showing that all the subprinciples have been fully evolutionally developed.

in all its seven aspects. *The Secret Doctrine* expresses it:

> . . . every "Round" brings about a new development and even an entire change in the mental, psychic, spiritual and physical constitution of man, all these principles evoluting on an ever ascending scale.[14]

Here is the description of the evolutionary phase of the first four principles during the four Rounds. The first principle, Sthūla-sarīra—the physical vehicle—in all its seven aspects undergoes its evolutionary development during the First Round. The Second Round continues the evolutionary development of the second principle, Linga-sarīra—the model body—in its seven aspects. The Third Round was concerned with the evolutionary development of the third principle, Prāna—the vital principle—in all its seven aspects. The Fourth Round carries on the evolutionary development of Kāma—the desire principle. The present evolutionary development of humanity is in the Fourth Round and is performing its fifth major developmental stage of evolution—termed the Fifth Root-Race. The specific evolutionary phase that is being developed now is the mānasic (or fifth) aspect of the fourth principle, Kāma. The sixth major evolutionary developmental phase of the Kāma principle will stress its buddhic aspect. The seventh major evolutionary development will be the unfoldment of the ātmic aspect of the Kāma principle, on this globe.

There is yet another way in which the forty-nine Fires are referred to, also in a Commentary from the Book of Dzyan:

> So also with man every "man [every principle] in man."[15]

The clue to the understanding of this apocalyptic sentence is provided in two words added by H. P. Blavatsky: "every principle." Since man is composed of seven principles, the sentence reiterates that each one of his principles

---

14. Ibid. I:162; I:216, 6-vol. ed.; I:185, 3rd ed.

15. Ibid. II:29; III:42, 6-vol. ed.; II:32, 3rd ed.

is subdivided into seven corresponding aspects. The list below shows the seven aspects for the principle of Ātman, and each one of the six remaining principles may be listed in the same manner:

Sthūla-śarīra aspect of Ātman
Linga-śarīra aspect of Ātman
Prāna aspect of Ātman
Kāma aspect of Ātman
Manas aspect of Ātman
Buddhi aspect of Ātman
Ātman aspect of Ātman

In *The Mahatma Letters to A. P. Sinnett,* reference is made to "the seven within the seven" as "sub-principles":

> Every element having its seven principles and every principle its seven sub-principles . . .[16]

This is another manner of expressing the idea of the forty-nine Fires. The phrase in the Commentary—"So also with man and every 'man' in man"—brought forward the idea of the forty-nine Fires, linking up with another concept which relates "every 'man' in man" to the seven sacred planets; for it is stated that each one of man's seven principles "receives its light and life from its own especial Dhyāni."[17] The Dhyāni here signifies a Planetary Regent. The concept is further elaborated in the next sentence of the Commentary:

> Each [principle] gets its specific quality from its primary (the planetary spirit), therefore every man is a septenate (or a combination of principles, each having its origin in a quality of that special Dhyāni). Every active power or force of the Earth comes to her from one of the seven Lords.[18]

The seven Lords are the seven Planetary Regents.

---

16. Letter No. XV, p. 92, 2nd ed.; p. 90, 3rd ed.

17. SD II:29; III:42, 6-vol. ed.; II:32, 3rd ed.

18. Ibid.

How Man Is Completed

The seven principles are yet again mentioned symbolically in the Commentaries in connection with the subject of the completion of man. Reference is first made to the forms (rūpas) which were produced by the Lunar Pitris, here termed the Lha:

> Into the forms projected by the Lha (Pitris) the two letters (the Monad, called also "the Double Dragon") descend from the spheres of expectation.[19]

The "two letters" obviously signify the Monad, or the Dougle Dragon as stated. An additional note further explains:

> In the esoteric system the seven principles in man are represented by seven letters. The first two are *more* sacred than the four letters of the Tetragrammaton.[20]

The Tetragrammaton is a Greek compound word signifying the four letters used in Hebrew texts to represent the name of God, the Ineffable Name. By different transliterations of the Hebrew letters there are four modes: JHVH, JHWH, YHVH, or YHWH. The transliteration *Jehovah* is the result of a combination of the Tetragrammaton with the vowel points of Adonai (meaning "my Lord"). When the texts are read, Adonai is substituted for the Ineffable Name.

The Pythagoreans represented the "sacred four" by the Tetraktys, the triangle within the square. The actual word *tetraktys* is Greek for the number four. In the triangle within the square, the Unmanifested Logos is at the apex because the triangle itself signifies the three Logoi—the Second Logos and the Third Logos being emanated from the First Logos. The completion of the triangle forms the tetraktys. This is done by placing the Point above the triangle, thus

---

19. Ibid. II 57; III:67, 6-vol. ed.; II:60, 3rd ed. The significance of the "spheres of expectation" is given following the exposition of the two and four letters.

20. Ibid. II:57 fn.; III:67, 6-vol. ed.; II:60, 3rd ed.

forming the tetrad, which is also represented by the triangle within the square. The Point is "the eternally Self-Existent One (Eternal in *Essence,* note well, not in manifestation, and distinct from the Universal ONE."[21] The Universal One represents the Unmanifested Logos, from which the cosmos will be produced when the Second Logos and the Third Logos are emanated.

As for the "seven letters," neither the first two nor the other five letters are given; instead Sanskrit terms are employed in *The Secret Doctrine.*

Another note tells where the monads were stationed while waiting in preparation for incarnation on the Earth. They were resting in "the spheres of expectation" which are:

> The intermediate spheres, wherein the Monads, which have not reached Nirvāna, are said to slumber in unconscious inactivity between the Manvantaras.[22]

Here the sphere of expectation is Globe C of the Earth Chain. From there the monads transferred to Globe D and accomplished the peopling of our Earth.

The Commentary continues by relating that even though the Lunar Pitris had projected satisfactory rūpas for the monad (Ātma-Buddhi) the vehicles thus supplied were not ready to enable the monads to function self-consciously. A symbolic form is used to express the situation:

> In the highest worlds, the three are one, on Earth (at first) the one becomes two. They are like the two (side) lines of a triangle that has lost its bottom line which is the third fire. (Catechism, Book III, sec. 9)[23]

A note is added after the words "the three are one" signifying:

> Ātman, Buddhi, and Manas. In Devachan the higher element of the Manas is needed to make it a state of

---

21. Ibid I:88; I:152, 6-vol. ed.; I:116, 3rd ed.

22. Ibid. II:57 fn.; III:67, 6-vol. ed.; II:60, 3rd ed.

23. Ibid. II:57; III:68, 6-vol. ed.; II:60, 3rd ed.

> perception and consciousness for the disembodied Monad.[24]

The statement in the Commentary that the three are as one in the *highest worlds* indicates that during the devachanic interlude the monad is making the ascent to the superior globes of the Earth Chain. On accomplishing this ascension the monad is enabled to pursue the Outer Rounds. Observe that the closing words of the first sentence of the Commentary—"on Earth (at first) the one becomes two"—signify that Ātman becomes two—Ātma-Buddhi. Then the three last words of the second sentence—"the third fire"—signify Manas, which adds the bottom line to the two sides of the triangle, thus completing the triform figure.

In conclusion. A Commentary from the Book of Dzyan shows how man was completed:

> It is those who have deserted the Superior Spheres, the Gods of Will, who complete the Manu of illusion. For the "Double Dragon" has no hold upon the mere form. It is like the breeze where there is no tree or branch to receive and harbor it. It cannot affect the form where there is no agent of transmission (Manas, "Mind") and the form knows it not.[25]

This is indeed a remarkable statement: it signifies that when there is no consciously functioning mind-principle there is no binding force to hold the "Double Dragon" (the monad) to its fixed abode (the Earth). The monad is then free to pursue *its* circuit, to its Source from which it originated, just as the Gnostics, the ancient Persians, Egyptians, and Chaldeans described it in their terms, during its journey in the after-death states.

When the monad "has no hold upon the mere form" (the lower quaternary), without Manas functioning as "the agent of transmission," and so long as the mind-principle is dormant, unawakened, man is considered to be incomplete—"a Manu of illusion." Superior Beings from Superior Spheres

---

24. Ibid.

25. Ibid. II:57; III:67-8, 6-vol. ed.; II:60, 3rd ed.

were required to bring man to completion, even those superior to the Lunar Pitris who first provided assistance by projecting their rūpas. The task had to be accomplished by beings who had themselves consciously and successfully passed through the stage of complete mānasic evolutionary development, such as the Gods of Will—the sons of Mahat.

What is the significance of the "Superior Spheres?" They represent the superior globes of the Earth planetary system, which are declared to be "deserted" when the Gods of Will descend from them to assist the people of the Earth by awakening the mind-principle.

The Gods of Will are the Mānasaputras, who are also called the Agnishvātta Pitris and Lords of the Flame. It has been suggested that the Sons of Fire, or the Lords of the Flame, came from Venus. But the Commentary from the Book of Dzyan indicates that the Gods of Will—who are the Lords of the Flame—descended from Superior Spheres in order to assist humanity; and *The Secret Doctrine* also affirms this in the following passages:

> It is those who have deserted the Superior Spheres, the Gods of Will, who complete the Manu of illusion. (Commentary from the Book of Dzyan)[26]
>
> Therefore, when the hour strikes for Purusha to mount on Prakriti's shoulders for the formation of the Perfect Man—rudimentary man of the first two and a half Races being only the *first,* gradually evolving into the most perfect of mammals—the Celestial "Ancestors" (Entities from preceding worlds, called in India the Śishta) step in on this our plane, as the Pitris had stepped in before them for the formation of the physical or animal-man, and incarnate in the latter.[27]

Attention is directed to the statement that the Celestial Ancestors—signifying the Lords of the Flame—are regarded as "Entities from *preceding worlds*" and that they simply *step in* on this our plane.

---

26. Ibid. II:57; III:67-8, 6-vol. ed.; II:60, 3rd ed.

27. Ibid. I:248; I:293, 6-vol. ed.; I:268, 3rd ed.

In giving an explanation of Stanza III, Śloka 13: ("The Lords of the Flame remain behind. They would not go, they would not create,") H. P. Blavatsky asked the question: "Who are the Lords of the Flame?" and then supplied the answer:

> [They are] the *Agnishvātta* who, as shown in Śloka 13, "remain behind" instead of going along with the others to create men on Earth. But the true esoteric meaning is that most of them were destined to incarnate as the *Egos* of the forthcoming crop of Mankind.[28]
>
> The Endowers of man with his conscious, immortal Ego, are the "Solar Angels"—whether so regarded metaphorically or literally. The mysteries of the Conscious Ego or human Soul are great . . . trans-Himalayan Occultists regard them as evidently identical with those who in India are termed *Kumāras, Agnishvāttas*. . . .[29]

Another name for the Lords of the Flame is "Fire Dhyānis" and once again they are identified with the Agnishvātta Pitris, the Solar Pitris:

> That class of the "Fire Dhyānis," which we identify on undeniable grounds with the Agnishvāttas, is called in our school the "Heart" of the Dhyān-Chohanic Body; and it is said to have incarnated in the third race of men and made them perfect. The esoteric Mystagogy speaks of the mysterious relation existing between the hebdomadic essence or substance of this angelic Heart and that of man. . . .[30]
>
> Therefore, the *FLAMES,* whose functions are confused in the exoteric books, and who are called indifferently Prajāpati, Pitris, Manus, Asuras, Rishis, Kumāras, etc., etc., are said to incarnate personally in the Third Root-Race and thus find themselves "reborn over and over again." In the Esoteric doctrine they are generally named the Asuras, or the *Asu-ra Devatā* or *Pitar*-devatā (gods) for, as said, they were first Gods—and the highest—before they became "*no*-gods," and had from Spirits of

---

28. Ibid. II:79; III:89, 6-vol. ed.; II:83, 3rd ed.

29. Ibid. II:88; III 97, 6-vol. ed.; II:92, 3rd ed.

30. Ibid. II:91; III:100, 6-vol. ed.; II:96, 3rd ed.

> Heaven fallen into Spirits of the Earth—*exoterically,* note well, in orthodox dogma.[31]

On this theme, a note was supplied in regard to the Gods of Will: "The *Suras,* who become later the A-Suras."[32] This may be interpreted in more ways than one. For instance, because the Suras (gods) deserted the Superior Spheres they became A-Suras (not-gods) ; or because some of the Suras "took in man their abode" (quoting a Commentary) they were gods no longer. As to why the Asuras came to be regarded as "demons" in exoteric literature need not be discussed here. However, the root meaning of the word *sura* (which signifies a god, a divinity) is interesting. Some state it is derived from the verbal root *svar,* to shine, hence a shining being. Some suggest that it comes from the word for the sun *sūrya,* the shining one. In any event, the Gods of Will signify the Mānasaputras, who awakened the fire of mind in man. The point is that because of having arrived at the stage of humanity on a previous Ladder of Life and graduated therefrom to continue their evolutionary development on superior stages, the Mānasaputras were able to awaken in man the mind principle, so that he was no longer a "Manu of illusion." He may now function as a true Manu, a thinking being. However, this did not take place until the third major evolutionary developmental stage of humanity, which occurred during the period of the Third Root Race, many ages after the actual peopling of the Earth had taken place.

The Secret Doctrine identifies the Gods of Will with the Mind-born Sons—Mānasaputras:

> . . . the hosts of these Sons of Light and "Mind-born Sons" of the first manifested Ray of the UNKNOWN ALL, are the very root of spiritual man.
>
> The Doctrine teaches that in order to become a divine, fully conscious god—aye, even the highest—the Spiritual primeval INTELLIGENCES must pass through the human

31. Ibid. II:247-8; III:250, 6-vol. ed.; II:258, 3rd ed.

32. Ibid. II:57; III:68, 6-vol. ed.; II:60, 3rd ed.

> stage. And when we say human, this does not apply merely to our terrestrial humanity, but to the mortals that inhabit any world, i.e., to those Intelligences that have reached the appropriate equilibrium between matter and spirit, as *we* have now, since the middle point of the Fourth Root-Race of the Fourth Round was passed. Each Entity must have won for itself the right of becoming divine, through self-experience. . . . The Mind-born Sons, the Rishis, the Builders, etc., were all men—of whatever forms and shapes—in other worlds and the preceding Manvantaras.[33]

Since other worlds and *preceding* Manvantaras (earlier world-periods of planetary activity) are mentioned, we may assume that reference is made to planetary systems *(plural)* prior to the Moon Chain. Had the system of which the Moon is the fourth globe of the Moon Chain been implicated, it would surely have been referred to as the *former* system *(singular)* of worlds. It was the Moon system of globes that bore and nurtured the Pitris, the Lunar Fathers, and in its turn passed them on for the peopling of the Earth.

---

33. Ibid. I:106-7; I:167, 6-vol. ed.; I:132, 3rd ed.

16.

CHRONOLOGY FOR THE PEOPLING OF THE EARTH

THERE is one subject which thus far has not been considered—that of the chronology connected with the peopling of the Earth. When did it occur? Does the Esoteric Philosophy consider this factor? Since it postulates that law and order prevail throughout a system, whether it be terrestrial, lunar, planetary, or solar, obviously time is essential for carrying on the necessary activities. One of the basic postulates of the Esoteric Philosophy is that for every period of activity there is a consequent period of rest. This signifies the law of periodicity, and time-periods are governed accordingly. On Earth, our day-to-day activities could not be carried on without a systematized observance of the time factor, and this is evidenced by the sequence of days and nights. Stated scientifically, this time factor is expressed by the equinoxes—the twice yearly periods of equal days and nights of twelve hours each. This then becomes the basic factor for the calculation of time: a new day every twenty-four hours.

The Significance of the Yugas

The chronological time-period relating to the Earth is conveyed in the Hindu scriptures by means of the sequence of days and nights. A Day of Brahmā is regarded as the time-period of the existence of the Earth covering its period of activity, whereas a Night of Brahmā relates to a period of dissolution. In the *Bhagavad-Gītā* a Day of Brahmā is stated to be equivalent to 1,000 revolutions of the yugas—in this case signifying *Mahāyugas.* Therefore, by determining the time-period of the yugas one may calculate the length of the period of activity of the Earth, or its lifetime. The Sanskrit word *yuga* signifies a period of time, or a cycle.

The basic factor for the cycle of the yugas is represented by the numbers 1, 2, 3, 4, which are applicable to the four yugas named Kaliyuga, Dvāparayuga, Tretāyuga, Kritayuga (or Satyayuga). The key factor in this series of numbers is 1,000. The cycle represented by the key figure of 1,000 is the Kaliyuga; the word *kali* gives the clue to its significance, for it means the die with the one spot, or number 1. Another important feature enters into the consideration of the yugas. Each of the four cyclic periods is introduced by means of a Sandhyā and is concluded by a Sandhyānsa. Both Sandhyā and Sandhyānsa mean a juncture, or a union. especially the juncture of day and night, and the juncture of night and day. Sandhyā is usually applied to the Dawn—the introductory period to yuga—represented as one-tenth of the period of the yuga. Likewise Sandhyānsa is regarded as the Twilight or concluding portion of the yuga, also equivalent to one-tenth of its length. Thus the total length of the Kaliyuga is described as lasting 1,200 *divine years.* This period is equated with "earth years" by multiplying 1,200 divine years by 360—resulting in 432,000 years (360 is the key number of the circle, credited to the Chaldeans, who divided the circle into 360 degrees).

The next cyclic period is the Dvāparayuga. Dvāpara is a word which means the die with the two dots; consequently it represents the means for denoting the time-period of the second yuga—2,000. With the addition of the Sandhyā and Sandhyānsa, the Dvāparayuga's time-period is 2,400 divine years or 864,000 earth years.

The key word for the third cycle of time—Tretāyuga—conveys its significance, for *tretā* means the die with the three dots. Consequently 3,000 with its addition of the Sandhyā and Sandhyānsa brings the total to 3,600 divine years, which is equivalent to 1,296,000 earth years.

The fourth yuga is known as the Kritayuga. Krita means well done, good, and the yuga is regarded as the Golden Age. It is also called the Satyayuga, *satya* signifying truth—

the time-period during which truth reigns supreme. Its key number is four, so that its cycle of time is 4,000 yugas which, with its Sandhyā and Sandhyānsa, totals 4,800 divine years, or the equivalent of 1,728,000 earth years.

The significance of the term Mahāyuga now becomes apparent. The Sanskrit compound term means "the great cycle" and represents the sum of the four yugas: 12,000 divine years or 4,320,000 earth years. This serves to establish the time-period of a Day of Brahmā as consisting of 4,320,000,000 years—equivalent to 1,000 Mahāyugas.

By using the Chaldean system of reckoning, based on the key figure 666, one is enabled to arrive at the same time-period. This is obtained by means of the two cycles known as the Saros and the Naros. The Chaldean cycle of the Saros consists of 120 years; this figure is obtained by doubling the 6 in the second position of 666; i.e. $60+60=120$. The cycle of the Naros is represented by the 6 in the first position of 666; i.e., 600. However, there were three cycles of the Naros termed the Greater, the Middle, and the Less. The Greater Naros is represented by the number 600,000. When this number is multiplied by the Saros, as well as by the key number in the second position of 666, the same figure is obtained as that given in the Hindu scriptures: $600{,}000 \times 120 \times 60 = 4{,}320{,}000{,}000$.

The Time-Period of the Human Period of Activity on the Earth

The epoch during which the peopling of the Earth took place is indeed recorded in *The Secret Doctrine,* being introduced by the statement that it is "vouched for by the learned Brahmans of Southern India" in the

> Tamil calendar called the "Tirukkanda Pañchanga," compiled, as we are told, from, and in full accordance with, secret fragments of Asuramaya's data.[1]

---

1. SD, II·67; III:77-8, 6-vol. ed.; II:72, 3rd ed. "Asuramaya is said to have been the greatest astronomer of the islands of Atlantis." (Ibid.) The Tirukkanda Pañchanga "was compiled by two very learned Brahmans"—"Chintamany Raghanaracharya, son of the famous Government astronomer of Madras, and Tartakamala Venkata Krishna Rao." (SD II:51; III:62, 6-vol. ed.; II:72, 3rd ed.)

Here is the statement from the calendar:

> The number that elapsed since the "Vaivasvata-Manvantara"—or the human period—up to the year 1887, is just 18,618,728 years.[2]

This signifies that for the year 1975, the beginning of the human Life-wave on this Earth (Globe D) took place 18,618,816 years ago.

The term Vaivasvata-Manvantara derives from the tradition that Vaivasvata represents the Fourth Round Root Manu—that is to say the Manu who acts as the Regent or Watcher of the seven globes of the Earth system during their periods of activity. As for the word Manvantara, although literally signifying the period "between Manus," it means here a period of activity, so that in this case the Vaivasvata-Manvantara has reference to the specific present Fourth Round period of activity on our Earth.

As to the specified number of years, one wonders how this figure was obtained. This information was not supplied in *The Secret Doctrine,* nor were the time-periods given in regard to the Root Races which preceded our present Fifth Root Race. Is the figure of 18 million odd years to be divided equally between the preceding Races? And what about the subraces which existed prior to the present subrace? Of greater importance would seem the length of time that the human race will have to carry on its activity on this sphere (Globe D). This figure is obtainable. It was given in an article written by H. P. Blavatsky although not published during her lifetime. It is preserved in the archives of The Theosophical Society at Adyar. This significant document was published in the March 1938 issue of *The Theosophist* and was prefaced by this notation, prepared by Boris de Zirkoff under the title: "On Cosmic Cycles, Manvantaras and Rounds":

> The MS. of this unfinished essay, in H.P.B.'s handwriting, exists in the Adyar Archives. Some of its pages are missing, and some of the sentences are broken off.

---

2. Ibid. II:69; III:79, 6-vol. ed.; II:72, 3rd ed.

> There is no definite clue in it which would help to determine the date at which it was written, except for the fact that a footnote mentions the sixth and seventh editions of *Isis Unveiled.* This MS. contains numerical relations and data not mentioned by H.P.B. anywhere else in her writings. It contains important keys which some students might be able to apply to various cosmological problems arising in their individual studies. The most noteworthy point in connection with this MS. is that it is written in two different handwritings, one of which is larger and more rounded than H.P.B.'s ordinary one.[3]

In view of the fact that the following statement is made in H. P. B.'s hand-written document, one comment may be added to the above notation:

> The astrological work states, that: "The number of years that elapsed since the beginning of Vaivasvata Manvantara—equals 18,618,725 years."[4]

Since *The Secret Doctrine's* figure is given as 18,618,728 for the year 1887, this shows that H. P. B.'s article was penned in 1884. Her document also supplied the information needed to solve the above-mentioned problems.

Before supplying the figures of the time-periods of the Root Races, the following statement from H. P. Blavatsky's article determines *how* the figures of the time-periods are calculated. It should be borne in mind that at the time this article was penned, H. P. B. used the term *"planet"* for each one of the seven spheres (or members) of the Earth planetary system. In *The Secret Doctrine* the seven members are referred to as Globe A, Globe B, Globe C, Globe D (our Earth), Globe E, Globe F, Globe G.

> Till now, the exoteric works gave only the period of the day of *Brahmā,* without either noting the other periods which might help toward the discovery of the secret, or giving that key itself which might provide the results now shown. But if we have the period of the Day of

3. *The Theosophist* (Adyar), issue of March, 1958, Vol. 59, p. 367.

4. *The Theosophist,* Vol. 59, p. 368.

> *Brahmā* and if we know that there are seven rounds, that each round covers seven planets, that on each planet there are seven races, and that the period of rest of a planet in every round equals that of its activity, and if to all this knowledge we apply the key of the septenary arithmetical progression series, then we get the numbers as given above [in this article]. There is a gradual rise of from one to seven. The duration of the existence of humanity during the Seven Rounds is 1: 2: 3: 4: 5: 6: 7. In each Round, the duration of the existence of humanity, on the seven planets of our chain is 1: 2: 3: 4: 5: 6: 7. The period of human existence in seven races, on one planet, is again 1: 2: 3: 4: 5: 6: 7. Now, as the planet evolves the 7 races in succession, before humanity can pass on to the next planet, the interval between the disappearance of humanity from one planet and its reappearance on the next, is equal to its existence on the planet which it has just left. Take then 4,320 millions as the day of *Brahmā,* and calculate according to the above explanation and you will arrive at the above given results.[5]

The process for the calculation of the figures was outlined above. Attention is first directed to the figures which give the total length of time required for humanity to accomplish its seven stages of evolutionary development—or seven Root Races—on our Earth (Globe D), here termed "the Human Period of Activity" in this Fourth Round. It is stated in this manner:

> Now, it will be seen that 44,081,632 years is the Human Period of Activity of our Planet in this Round.[6]

Since the Human Period of Activity on this Globe has occupied 18,618,816 years (up to 1975), it can be seen that the halfway mark of the Human Period of Activity has not been reached, for there are still 25,462,816 years to be experienced on the Earth before the Human Life-Wave moves on to Globe E. This shows also that the time-periods of the

---

5. Ibid. Vol. 59, pp. 370-1. "The above given results" signify the figures of the Rounds, Globes, and Root Races given in the document. These will be presented here in due course.

6. Ibid. Vol. 59, p. 370.

early Root-Races were considerably shorter than will be those of the last three Root-Races. Here are the time-periods of the first four stages of evolutionary development—termed the Four Root-Races—on our Earth:

Duration of the		
First Root-Race	1,574,344	Years
Second Root-Race	3,148,688	
Third Root-Race	4,723,032	
Fourth Root-Race	6,297,376	
Duration of Four Root-Races up to the fifth:	15,743,440	Years
Number of Years "lived" by the fifth Root-Race	2,875,285	Years (up to 1884)
	18,618,725	Years—"the number

of years elapsed since the beginning of Vaivasvata Manvantara." For the number of years up to 1975, add 91 years, giving the total of 18,618,816 years.

Following is the tabulation of the Seven Root-Races, representing seven stages of evolutionary development which will be accomplished by humanity during its period of activity on this Earth:

Duration of the		
First Root-Race	1,574,344	Years
Second Root-Race	3,148,688	
Third Root-Race	4,723,032	
Fourth Root-Race	6,297,376	
Fifth Root-Race	7,871,720	
Sixth Root-Race	9,446,064	
Seventh Root-Race	11,020,408	
	44,081,632	Years (on Globe D)[7]

The Time-Period of Human Activity on the Globes of the Earth System During the Fourth Round

In addition to the length of time required for the evolution of the Seven Root-Races on Globe D, the number of years required for the human period of activity on each

7. Ibid. Vol. 59, p. 370.

one of the seven Globes of the Earth Planetary System is tabulated. The figures represent the accomplishment of seven stages of evolutionary development on each one of the Globes of the planetary system, comparable to the evolution of Seven Root Races in process on Globe D. At the present time, the Fifth Root Race represents the fifth stage of evolutionary development. Here is the tabulation:

Duration of Humanity on Globe A	11,020,408	Years
Globe B	22,040,816	
Globe C	33,061,224	
Globe D	44,081,632	
Globe E	55,102,040	
Globe F	66,122,448	
Globe G	77,142,856	
	308,571,414	Years[8]

The Time-Period of the Rounds

The next time-period to be considered is that which deals with the passage of humanity on the circuit of all the members of the Earth planetary system, which is referred to as the passage on the seven Globes. One complete circuit of the seven globes has been termed One Round, for it represents the passage of the human Life-Wave on each one of the seven globes in serial order.

Duration of the First Round	154,285,714	Years
Second Round	308,571,428	
Third Round	462,857,142	
Fourth Round	617,142,856	
Fifth Round	771,428,570	
Sixth Round	925,714,284	
Seventh Round	1,079,999,998	
	4,319,999,992	Years[9]

It is noticeable that the total figure here given for the Seven Rounds is 8 years less than the Brahmanical period of 4,320,000,000 years for the Day of Brahmā. H. P. Blavatsky

8. Ibid. Vol. 59, p. 370.

9. Ibid. Vol. 59, p. 369.

explained the reason for this in a footnote:

> It will be obvious that for the purpose of having round numbers, we have, in our calculations, omitted fractions. Thus on the whole "Day of Brahmā" we have left off a period of eight years. It should also be noted that each "Round" period in the above table signifies both the period of planetary Activity and interplanetary Rest.[10]

The interplanetary rest signifies that during the process of humanity's passage from globe to globe of the Earth system, after the human Life-Wave has completed its seven stages of evolutionary development on one globe, a period of rest takes place before it passes on to the next globe.

Attention is directed to the fact that in the tabulation of the time-periods of the Seven Rounds, the Fourth Round is listed as lasting for 617,142,856 years, whereas in the tabulation of the time-periods of humanity on each globe of the system for the Fourth Round, the total figure in that tabulation was given as 308,571,414 years. This is but half the total number of the figures in the Round tabulation. This difference was explained by H. P. Blavatsky when referring to the Round tabulation. In the quotation which follows, the period of a Kalpa refers to a Day of Brahmā. Then, too, the time-period of a Satyayuga is 1,728,000 years.

> We have thus 617,142,856 years as the period of our Fourth Round. And as the "Night of Brahmā" or period of Rest, is always equal to the "Day of Brahmā" or the period of activity on each planet,—the period of activity in this Fourth Round equals 308,571,428 years. It thus exceeds the period of duration given for our Manvantara (308,448,000 years) [i.e. our Fourth Round] in the Brahmanical calculations, only by 123,428 years; and this would be made away with, if in making this calculation we had deducted from it the overlapping of the period of Kalpa which is equivalent to one Satyayuga and which the Brahmans for purposes of esoteric secrecy have added to the "Day of Brahmā."[11]

10. Ibid.

11. Ibid. Vol. 59, p. 369.

There is yet another item to take into consideration. It is the difference in the figures of the Fourth Round as just quoted in the above paragraph, which were given as 308, 571,428 years. In the tabulation of the time-period of activity on each globe of the system the total figure is listed as 308, 571,414 years—a difference of 14 years. Again H. P. Blavatsky explained this difference:

> For the sake of having round numbers, we are again obliged to leave off fractions and hence there is a slight difference. This figure when doubled, will give [617, 142,828 which is] 28 years less than the 4th Round period mentioned above [617,142,856]. Here in the period of activity we have a difference of only fourteen years.[12]

This concludes the presentation of the time-periods supplied in this significant document unpublished during H. P. Blavatsky's lifetime; but her closing words should be read at this point:

> It is noteworthy that in the Hindu exoteric works the period of the Manvantara (One Round) is given at 308 millions, to speak in round numbers. Now two reasons may be assigned for the adoption of that course. In the first place, the duration of the 4th Round according to the above calculations is 617 millions again, to use a round figure. Now, we have already stated that the period of activity of the planetary chain in one round is equal to its period of rest during the same round, while humanity rests in its passage from planet to planet. Thus divide the period of the 4th Round, into two equal parts; and you have 308 millions and odd as the *Manvantaric* period of *our* Round. Thus our Round period may have been in the first instance taken as the *Manvantaric* period. The second reason may be this. Our planet being the exactly middle period and we being in the middle of the seven rounds, our round period may have been taken to denote the *average Manvantaric* period, thus at the same time giving a key in a

12. Ibid. Vol. 59, p. 370

veiled form to the mystery of the geometrical progression. We have already stated that the above figures are exact, if the exoteric calculations of the Brahmans about the day of Brahmā be correct. But we may again state here that that figure is not correctly given out in exoteric numbers. We may, however, add that the explanations given by us about the progressions, etc., are facts and can be faithfully utilized when any one of the above-described figures is correctly known—in calculating all the rest of the figures. And these processes we have explained because we know that not one of the *exact* numbers will ever be given out, as they pertain to the Mysteries of Initiations and to the Secrets of the occult influence of Numbers.[13]

The Time-Periods of the Subraces

Thus far no information has been imparted about the subraces, even though they are so frequently mentioned in *The Secret Doctrine.* The reason for this is simply because H. P. Blavatsky did not provide any instruction on the subject of the subraces in her article on the Rounds and Races.

The term subrace bears no connotation of a lower or inferior racial strain. It is intended to represent a division in the stage of evolutionary development, during which minor developmental aspects may be evolved. Thus seven subraces, numbered 1 to 7, are regarded as coming into existence during one major stage of evolutionary development.

While working on the numerical progressions given in H. P. Blavatsky's article, after several efforts I discovered that her system may likewise be applied to the subraces. As a consequence, the time-periods allocated to the subraces may be demonstrated to concur with the figures which have been presented by H. P. Blavatsky in her tabulation of the Seven Root-Races. Therefore the total period designated for subraces 1 to 7 inclusive, are precisely those given by H. P. Blavatsky for the First Root-Race in her article.

---

13. Ibid. Vol. 59, pp. 371-2.

DURATION OF THE SUBRACES DURING THE TIME-PERIOD OF THE FIRST ROOT RACE

First subrace	128,519	Years
Second subrace	160,648	
Third subrace	192,777	
Fourth subrace	224,906	
Fifth subrace	257,035	
Sixth subrace	289,165	
Seventh subrace	321,294	
Time-Period of	1,574,344	Years of First Root-Race

As H. P. Blavatsky pointed out when making her numerical progression it was necessary to indicate a slight difference in the figures; it was likewise necessary in the above tabulation to make a slight difference in the numerical tabulation, applicable only to the last figure of the sixth and seventh subraces—because of working with number 7. The same notation will apply to the last column of figures in all the other tabulations of the subraces to be presented, due to the same reason.

DURATION OF THE SUBRACES DURING THE TIME-PERIOD OF THE SECOND ROOT-RACE

First subrace	257,039	Years
Second subrace	321,298	
Third subrace	385,556	
Fourth subrace	449,808	
Fifth subrace	514,072	
Sixth subrace	578,329	
Seventh subrace	642,588	
Time-Period of	3,148,688	Years of Second Root Race

DURATION OF THE SUBRACES DURING THE TIME-PERIOD OF THE THIRD ROOT-RACE

First subrace	385,558	Years
Second subrace	481,945	
Third subrace	578,332	
Fourth subrace	674,719	
Fifth subrace	771,106	
Sixth subrace	867,493	
Seventh subrace	963,879	
Time-Period of	4,723,032	Years of Third Root Race

DURATION OF THE SUBRACES DURING THE TIME-PERIOD OF THE FOURTH ROOT-RACE

First subrace	514,077	Years
Second subrace	642,593	
Third subrace	771,109	
Fourth subrace	899,625	
Fifth subrace	1,028,141	
Sixth subrace	1,156,657	
Seventh subrace	1,285,174	
Time-Period of	6,297,376	Years of Fourth Root Race

DURATION OF THE SUBRACES DURING THE TIME-PERIOD OF THE FIFTH ROOT-RACE

First subrace	642,596	Years
Second subrace	803,241	
Third subrace	963,886	
Fourth subrace	1,124,531	
Fifth subrace	1,285,177	
Sixth subrace	1,445,822	
Seventh subrace	1,606,467	
Time-Period of	7,871,720	Years of Fifth Root Race

There is a question which has not been determined satisfactorily: in what subrace is the Fifth Root-Race now functioning? The next tabulation indicates that the time-period of the Fourth subrace has not concluded its term.

THE PRESENT SITUATION OF THE FIFTH ROOT-RACE IN RELATION TO ITS SUBRACES

(Time-period of the Fifth Root-Race: 7,871,720 years)

SUBRACES	(in years)		FIFTH ROOT-RACE
1st	642,596		Number of yrs. lived by 5th R-Race to 1975 (since 4th R-Race):
2nd	803,241		
3rd	963,886		2,875,376
	2,409,723	(Total, 1 to 3)	Required yrs. to complete 4th Subrace:
4th	1,124,531		658,878
	3,534,254	(Total, 1 to 4)	3,534,254
5th	1,285,177		
	4,819,431	(Total, 1 to 5)	Required yrs. to complete 5-6-7 Subraces:
6th	1,445,822		4,337,466
	6,265,253	(Total, 1 to 6)	Time-period of 5th Root-Race:
7th	1,606,467		
	7,871,720	(Total, 1 to 7)	7,871,720

DURATION OF THE SUBRACES DURING THE TIME-PERIOD OF THE SIXTH ROOT-RACE

First subrace	771,116	Years
Second subrace	963,889	
Third subrace	1,156,664	
Fourth subrace	1,349,438	
Fifth subrace	1,542,212	
Sixth subrace	1,734,986	
Seventh subrace	1,927,759	
Time-Period of	9,446,064	Years of Sixth Root-Race

DURATION OF THE SUBRACES DURING THE TIME-PERIOD OF THE SEVENTH ROOT-RACE

First subrace	899,635	Years
Second subrace	1,124,538	
Third subrace	1,349,440	
Fourth subrace	1,574,345	
Fifth subrace	1,799,248	
Sixth subrace	2,024,149	
Seventh subrace	2,249,053	
Time-Period of	11,020,408	Years of Seventh Root-Race

This concludes the chronological account of the time-period of activity for humanity on this Earth (Globe D). When the Seventh Root-Race will have concluded its time-period, the human period of activity on this globe will cease and the interplanetary period of rest will ensue. The human Life-Wave will then transfer its activities to the next globe (Globe E) of the Earth planetary system.

INDEX

ABBREVIATIONS

def.	defined (or definition supplied)
expl.	explanation
fn.	footnote
q.v.	quod vide (meaning "which see")
re	regarding
sig.	signification (or signifying)
Sans.	Sanskrit

QUEST BOOKS
are published by
The Theosophical Society in America,
a branch of a world organization
dedicated to the promotion of brotherhood and
the encouragement of the study of religion,
philosophy, and science, to the end that man may
better understand himself and his place in
the universe. The Society stands for complete
freedom of individual search and belief.
In the Theosophical Classics Series
well-known occult works are made
available in popular editions.